rad ROBOTS

rad ROBOTS

A celebration of awesome automatons: the mad, bad and dangerous to know

Kyle Cathie Limited

Simon Furman

For Anna,
my very own fembot.

First published in Great Britain in 2009
by Kyle Cathie Limited
122 Arlington Road
London NW1 7HP
general.enquiries@kyle-cathie.com
www.kylecathie.com

10 9 8 7 6 5 4 3 2 1

ISBN 978-1-85626-855-4

Cover picture acknowledgements:
Front cover, from left to right: Carolco/Kobal Collection; Advertising Archives;
Sci-Fi Channel/Kobal Collection.
Back cover, from left to right: AKG-images;
Paramount Pictures/Album/AKG-images;
MGM/Kobal Collection.

Project editor: Jennifer Wheatley
Designer: Jake Tilson Studio
Picture research: Julia Gelpke
Copy editor: Claire Rushbrook

A Cataloguing In Publication record for this title is available from the British Library.

Colour reproduction by Image Scanhouse

Printed and bound in China by SNP Leefung

Contents

Introduction

The existence of robots says a lot about the human condition, not least our psychological need/quest to recreate ourselves, biologically or otherwise.

Robots represent to us, with our limited lifespan, human frailties and penchant for waging war on one another, something of an ideal. Robots are, for the most part, created by man to be better than man. To be able to do the things we cannot, or do them better, faster or for longer. But therein lies the rub. For if robots are stronger and smarter than us, it stands to reason they will one day replace us entirely or reach a point in their evolution where they simply conclude—in entirely logical and dispassionate fashion—that humans are redundant, incapable of sustaining themselves as a race for any length of time. So while we embrace the idea of the robot, we also fear it. And maybe that's the real attraction.

After all, fear—as much as we might claim otherwise—is an attractive proposition. It reminds us we're alive. Fear reinforces the positives of our everyday existence.

Whenever we're reminded, however briefly, of our mortality, of our brief and tenuous hold on existence, the mundane and everyday facets of our lives take on a bright new gloss. Why else would we throw ourselves off bridges on an elasticated cord or down snow-covered mountains on two strips of fibreglass?

Robots—for all the significant and breathtaking leaps forward in the actual science of robotics and computers—remain, for now, the stuff of fantasy, whether in printed fiction, celluloid or small screen, and accordingly our unease remains remote, comfortable, something to enjoy with a tub of popcorn. But as the technological pace of the world increases and robots come nearer and nearer to a functional reality and eventually an everyday item, so our comfort zone narrows. Now, therefore, seemed like a good time to roll back through the entire history of robots, from the earliest automatons to today's cinematic CGI wonders, and consider what it is about the notion of a mechanical being that has so stirred the imaginations, dreams, aspirations and—oh yes—fears of so many.

The idea of a robot (though this exact term wouldn't be coined until 1921) or artificial being goes back a long way, all the way in fact to the myths and legends of the ancient Greeks. Their notion of a spark of life transferred to an inanimate human statue would eventually—via such works as Mary Shelley's *Frankenstein* and Karel Capek's *R.U.R.*—lead to today's radical, cutting edge robots. From then till now, all manner of mechanoids, of far-ranging pedigree and sophistication, have clanked, whirred and stamped their way into the public consciousness, pervading nearly every facet of popular culture, from books to movies to TV to advertising to toys to comic books. In their own way, and far more effectively than in any pulp sci-fi potboiler, robots have already conquered the world.

My own wide-eyed fascination with robots began as a kid. I vividly remember the horror and sci-fi comics published by Alan Class & Co (which, though I wouldn't learn this until much later, reprinted the American pre-Marvel Atlas Comics titles of the 50s), with lurid titles such as *Secrets of the Unknown, Uncanny Tales* and *Amazing Stories of Suspense*. The cheap, grainy paper stock and black and white presentation only added to the illicit thrill of stories bearing titles like 'I Became a Human Robot,' 'My Name is Robot X,' 'Elektro: He Held the World in his Iron Grip!', 'The Thing Called Metallo' and many, many more. Weaned on these tales, I sought out my robot fix in fiction, in particular pulp sci-fi magazines (featuring short stories from such authors as Isaac Asimov and Robert Heinlein) scavenged from second-hand emporiums such as Popular Book Centre and Plus Books, eyes carefully averted from "the back room," where—behind

RIGHT: *Tales of Suspense* #16, April 1961. Robots proliferated in the pulp-style multi-story comic books published by Atlas (who would later become Marvel) Comics. These tales of the weird and wonderful were reprinted in the UK by Alan Class & Co.

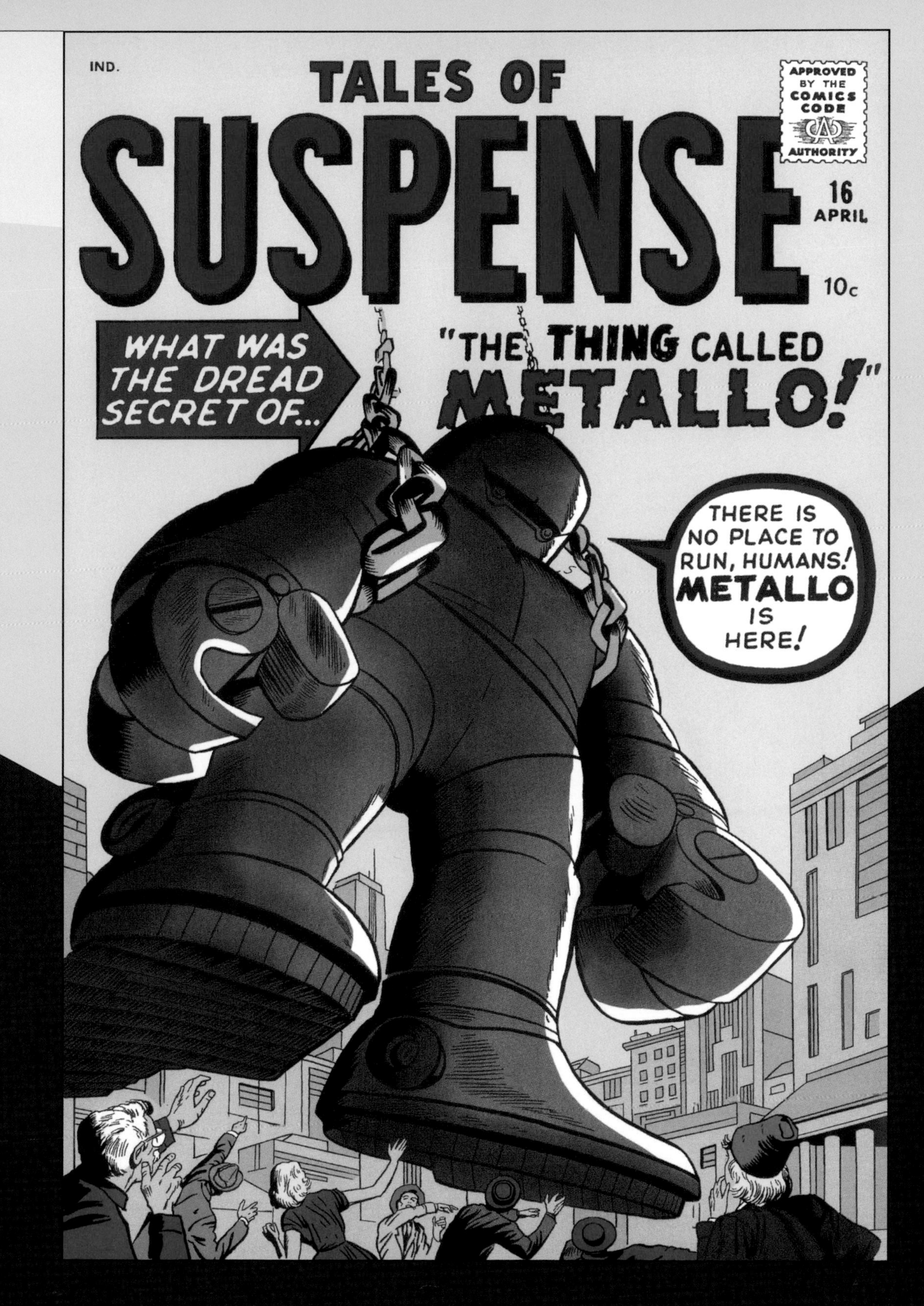

BELOW: poster for *Westworld*, the tale of a robotic theme park gone awry. Michael Crichton would later adapt the concept for his novel *Jurassic Park*. **OPPOSITE:** Cameron (a *Terminator* in-joke in itself), as played by Summer Glau in the TV series *Terminator: the Sarah Connor Chronicles*.

a beaded curtain—men leafed through soft porn magazines. Aged eleven, robots and aliens were my 'porn'.

Aged twelve I saw Michael Crichton's *Westworld*, a film that seared itself onto my brain, and has since lost little of its initial impact. Its depiction of a fantasy playground (the phrase 'theme park' meant nothing to me at the time) populated by robots, in which human visitors live out their fantasies was, to me, quite simply breathtaking. I still remember the poster tagline: "Westworld... where nothing possibly can go worng..." with the last few misspelled letters tumbling off the horizontal. I was gripped, scared and thrilled all at once. I had found my oeuvre!

Many mechs later, I was a twenty-something fledgling comics scriptwriter scrabbling around for work. Marvel UK (the London-based arm of American comics giant Marvel Comics) had recently acquired the comic book rights to *Transformers*, the "robots in disguise," and they were actively looking for writers to pitch story ideas. I knew nothing of either the toy line or accompanying animated TV series that, in 1985, were already spellbinding another generation of kids. It was simply a paying job and I was glad to have it. The fact that it was robots was neither here nor there. But whether I knew it or not or liked it or not, *Transformers* would come to shape and define my career as a writer, and now, some twenty-four years later, I'm *still* writing *Transformers* stories.

OPPOSITE: Nestor Class-5 robot Sonny from *I, Robot*, makes the grade for *Rad Robots*, whereas Del Spooner (Will Smith) with his mere cyborg implants doesn't. Got to draw the line somewhere!

Be that as it may, a lot of research or simply a refreshing of acquired knowledge was needed for this book. No onerous task either. I was more than happy to delve into both familiar and unfamiliar territory,

whether it was re-reading the Asimov robot stories or discovering *R.U.R.* (Rossum's Universal Robots), which I'm ashamed to say I'd never even come across before, or watching the magnificently remastered edition of Fritz Lang's 1927 movie *Metropolis* or catching up with the first season of *Terminator: The Sarah Connor Chronicles*. There are worse jobs.

Next, though, I needed to create a framework on which to hang what otherwise could turn out to be a sprawling, aimless (and merely mildly diverting) ramble through weird and wonderful robots on screen and printed page. I needed to define two things: what it was I was trying to say about robots and what exactly my working definition of a robot was. I tackled the latter proposition first. What is a robot? Is it merely any kind of artificial intelligence (A.I.) or does it need to have three dimensions... arms, legs, etc.? Do cyborgs (part human, part machine) and gynoids, a term for a female cyborg coined by British SF author Gwyneth Jones in her novel *Divine Endurance*, count? Should a robot, through necessity, be constructed of metal? Or does an artificial being fashioned from other materials, such as an Auton (animated showroom dummies from *Doctor Who*), constitute an automaton?

In the end, I decided pretty much everything should be included: metal robots, A.I.s (for example, HAL from *2001: A Space Odyssey* or the errant computer from *Demon Seed*), human robots (such as those in *R.U.R.* and even the Cylons from 'new' *Battlestar Galactica*), toy robots and real robots. In fact, the only sub-category I discarded entirely was cyborg/gynoids. So, for example, in *I, Robot* (the movie, rather than either the original short story by Eando Binder or the later amalgamation of Asimov's short stories on which it's very loosely based), while Sonny is a robot, the Will Smith character with his robotic implants is not.

This catchall decision soon started to give shape and form to the book itself. Rather than a purely chronological history of robots from whenever to the present day, I decided to group robots into loose amalgamations or classifications, evaluating them based on shared characteristics, aims or simply genre. (Though the first three chapters do progress things in a fairly linear fashion.) To each grouping I would apply my basic contention that what we create says more about the creator than the robot itself, that in the act of creating or coveting we are exposing our own ideals, desires and flaws. That's not to say that this book is in any way a lofty treatise on the human condition. It's not. It's a none-too-po-faced trawl through the many weird and wonderful and funny and terrifying robots to inhabit the many worlds created on screen and page (and beyond), a sort of personal journey on which you, the reader, are very much invited.

It's also a book you can dip in and out of. It doesn't necessarily require you to start at page one and finish at the acknowledgements. My main aim is to entertain and inform, and just perhaps provide some food for thought. Oh, and maybe—just maybe—to leave you with a slight edge of disquiet, a subliminal sense of anxiety at what exactly the future holds. Remember, fear is good.

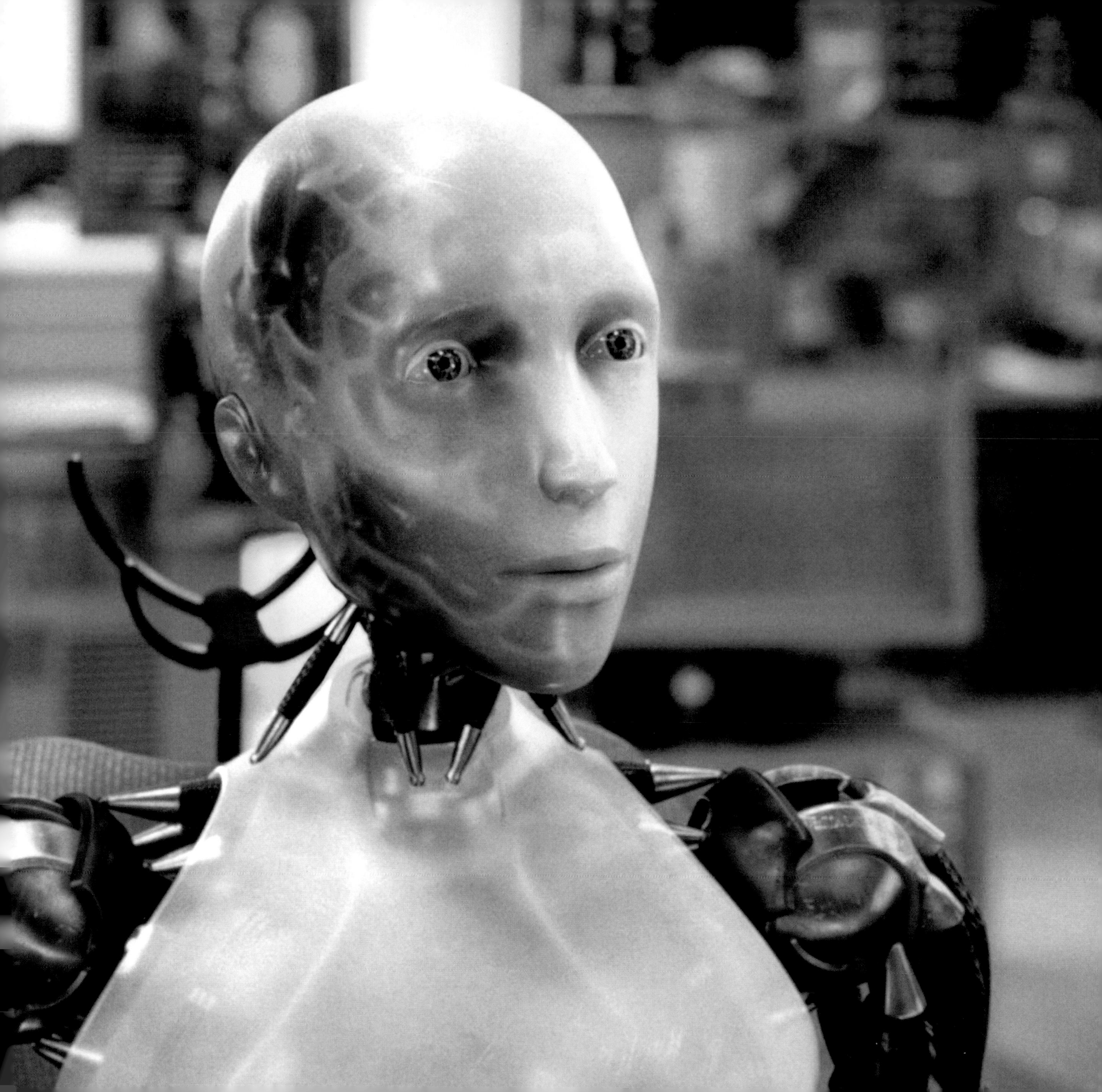

Automata to Robota, a mechanical evolution

1

Rise of the Robots

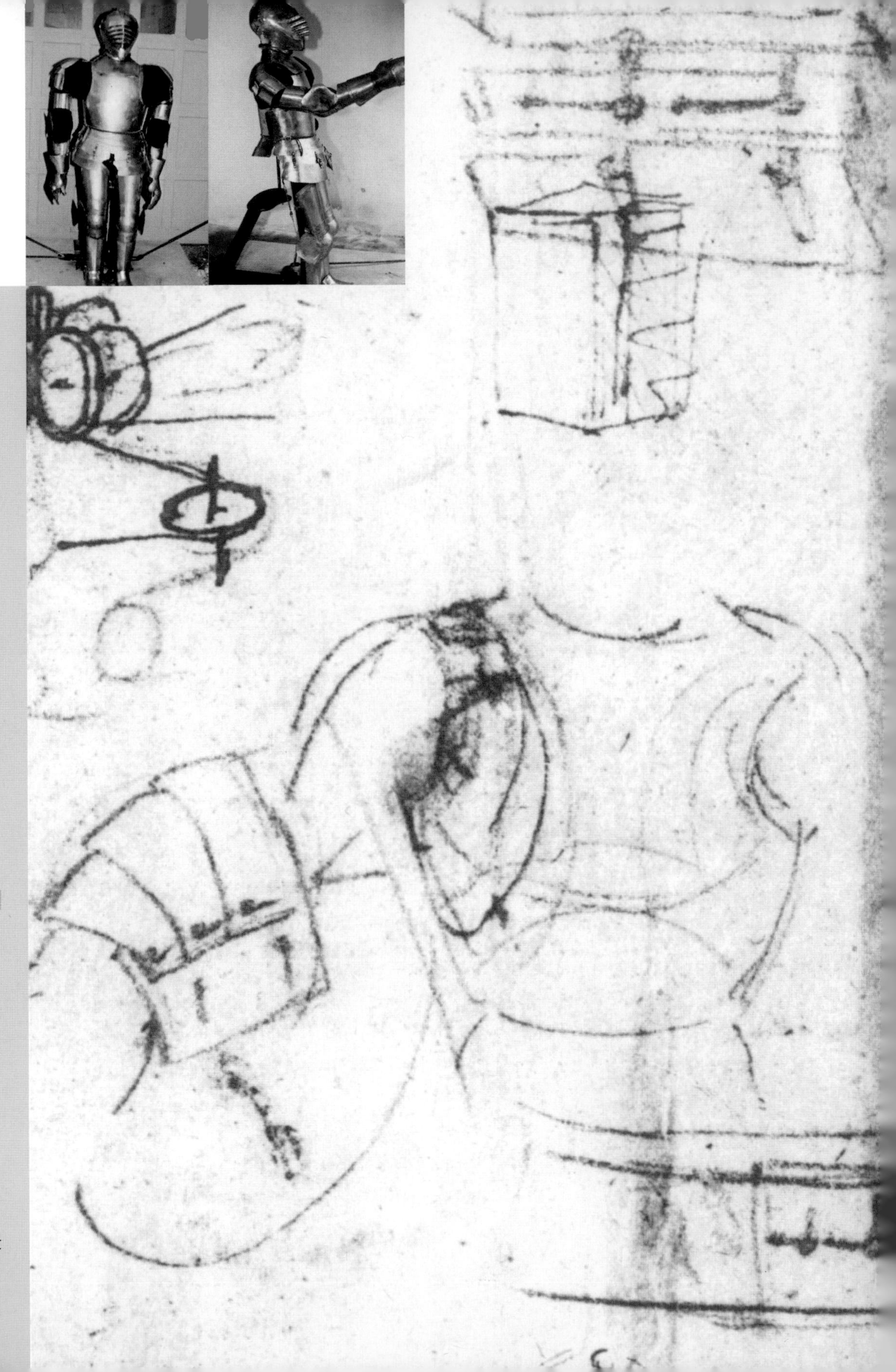

If I mention that-while not the first-Leonardo Da Vinci was at least one of the first to conceive of a mechanical man, I'm sure it will come as no great surprise to anyone with more than a passing interest in technological evolution. After all, the Renaissance man dreamed up a whole lot of stuff that wouldn't amount to actual scientific/mechanical achievement for another 400 years or so. Along with his many contributions to hydraulics, physics, mechanics and engineering, Leonardo devised, sketched and in some cases even built cranes, diving gear, helicopters, a parachute, land and water craft, tanks and missiles. Oh, and a robot.

Conceived and drafted after a period of his life spent extensively dissecting and analysing the human body, and springing forth from his earlier anatomical studies that led the famous Vitruvian Man, Leonardo's robot was essentially a suit of armour of German/Italian origin, with inner workings. It was designed in 1495, and while never actually built (to our knowledge) it was clearly intended to sit up, move its arms and neck and work an anatomically correct jaw. Robotics engineer Mark Roshiem (author of the award-winning *Leonardo's Lost Robots*) has since analysed and even built a prototype of Leonardo's robot

knight, based on the scant few preliminary sketches and notations that survived, and Leonardo's clear grasp of the fundamental principles of robotics is apparent.

It's been suggested that one source of inspiration for Leonardo's robot was the ancient Greeks, and it's among their myths and legends that the story of the robot—though that exact term wasn't coined until much later—really begins. The first mention of something approximating a robot, comes in the myths and Homeric poems relating to Hephaestus, Greek god of fire and the forge and master craftsman to the Gods. Depicted as crippled, misshapen and lame, Hephaestus' inner beauty was triumphantly represented through his many dazzling creations, such as Zeus's thunderbolts, Achilles' shield and Apollo's chariot. But it's the servants Hephaestus made for himself, cast from molten metal and imbued with the spark of life that begin our own, robotic odyssey.

Various texts describe the fruits of Hephaestus' labours. Tripods on golden wheels, supporting cauldrons of molten metal that moved around his workshop of their own accord, two golden handmaidens to help and support the crippled god and said to contain the accumulated wisdom and knowledge of the gods, making them—arguably—the first computers. As well as these 'robots', Hephaestus is also said to have crafted the giant bronze golem Talos, guardian of the isle of Crete, who repulsed Jason's ship, the *Argos,* by hurling giant boulders into the ocean, two fire-breathing mechanical horses presented to his sons, a giant eagle cast out of bronze to torment the chained giant Prometheus, sundry 'mechanical marvels' created for his lovely if routinely unfaithful wife, Aphrodite, and 'brazen-footed' mechanical bulls that breathed fire, presented as a gift to King Aeetes.

He was certainly a busy boy, Hephaestus, and even found time, at the behest of Zeus,

OPPOSITE: original Leonardo (Da Vinci) designs for his 'lost' robot knight, plus the 20th century realization of the knight, as built by robotics expert Mark Rosheim and featured in the BBC drama documentary *Leonardo*. Rosheim's robot knight is now owned by San Diegan Ben Sweeney.
ABOVE: "Tethys and Hephaestus" Roman fresco from Pompeii depicting the master craftsman to the Gods.

OPPOSITE: Talos, the giant bronze guardian of Crete, as visualized by legendary stop-motion effects supremo Ray Harryhausen in *Jason and the Argonauts.*

It wasn't just Hephaestus who was creating automata in the mythical mists of time

to mould/craft the first woman (Pandora, she of the box) from clay, or at least the Earth (Gaia). And while this latter example is stretching the definition of a robot somewhat, it plays neatly into the Jewish legend of the Golem, which we'll look at in more detail shortly, itself a precursor of what would become, if you like, the staple robot story. By the way, it's worth noting here that it was from Hephaestus that Prometheus stole fire, which also joins the dots in another storytelling loop. Bear with me here.

But it wasn't just Hephaestus who was creating automata, as they were collectively known, in the mythical mists of time. The Athenian craftsman Daedalus, who famously devised and built the labyrinth in Knossos in which the monstrous Minotaur was imprisoned, was also renowned for creating statues that moved. In fact, so the legend goes, Deadalus was indirectly responsible for the creation of the Minotaur he later helped imprison. It was he who fashioned a hollow metal or wood (the accounts differ) bull, so that King Minos' wife Pasiphae could seduce a real bull, the end result of which was the Minotaur. Pygmalion, a legendary inhabitant of Cyprus, carved a statue of a woman from ivory and promptly fell in love with it. Repeated prayers to the goddess Aphrodite (wife of Hephaestus, you see how these things interlink?) bear fruit when she takes pity on Pygmalion and brings the statue to life so they can marry. Those (ancient) Greeks!

All well and good, but did the ancient Greeks themselves actually build anything approximating a robot? There's certainly evidence that technology, as we understand it, existed well into BC. The Antikythera mechanism, an early astrological instrument/calculator, was discovered in 1900 in a wreck off the Greek island of the same name and dated to around 150–100 BC. It's essentially an analog computer with miniaturized parts and gears, though its exact function is still subject to speculation. Even so, it's evidence that the Greeks were highly advanced in technological terms. Similar artefacts would not feature for another thousand or so years! The island of Rhodes was well known in ancient Greece for automata. Pindar, one of the most famous of all Greek poets, writes of "animated figures..." in the streets that appear to "breathe in stone" and "move their marble feet."

Archytas of Tarentum was a Greek mathematician, statesman and philosopher who lived in southern Italy around 400 BC. He is sometimes referred to as the founder of mechanics, and is famous (in terms of what we today call robotics) for creating two mechanical devices, the first a bird, the second a rattle for children known as a 'clapper.' Of more interest to us is the bird,

ABOVE: un-credited French illustration of Archytas of Tarentum and his mechanical dove.

Literally a mechanical device designed to mislead and surprise

maybe modelled on a pigeon or a dove, which was constructed of wood and connected to a pulley and counterweight, so that when set in motion by a blast of steam or compressed air it would 'fly' from a lower perch to a higher perch, flapping its wings in the process. However, some academics believe there to have been a book on mechanical engineering already in circulation at that time by another Archytas altogether and that the creation of the mechanical bird is wrongly credited.

From these early beginnings, the fledgling science of mechanics and automata progressed, albeit slowly, through Chinese artisans who crafted mechanical men and even whole orchestras, via a noted Muslim alchemist and engineer of the 8th century, Jabir ibn Hayyan, whose Book of Stones included studies on artificial scorpions, snakes and humans, to Al-Jazari, an Arab scholar and inventor in the 13th century, whose book *Automata* provides a stunning insight into just how advanced technological thinking was in the east in the middle ages compared to the west. Unlike the Greek versions, early Arab automata seemed more for practical application than exhibition or whimsy, designed primarily to make everyday life easier. However, in common with other pre-industrial societies, these designs were destined to remain in the domain of the theoretical, rarely if ever (to our knowledge) turned into physical constructions.

All of which brings us back to Leonardo Da Vinci and his robot knight. Though loosely detailed on just a few sketches discovered in 1950s, it represents the closest equivalent to the 20th century notion of a mechanical man. Renaissance Europe embraced the idea of human automata with a prolific passion. First out of the gate, in around 1530, was German inventor Hans Bullman with a series of simulated people, often playing one musical instrument or another, and when in the 17th century philosopher René Descartes expounded his theory of dualism, which separated body and soul, with the former reduced to the functional level of a machine, it seemed to well and truly open the doors to all manner of intricate and lifelike human automata. French inventor Jacques de Vaucanson (1709–1782) produced some particularly fine examples of anatomically precise automata, including *The Flute Player* and *The Tambourine Player*.

In Japan, in the 18th Century (the Edo period), puppets called *karakuri-ningyo* featured internal mechanisms similar to the clock-making technology imported into the country by Francis Xavier and other Jesuit missionaries. The word *karakuri* implies hidden magic, literally a mechanical device designed to mislead and surprise.

But while human automata and puppets were certainly novel and, for the time,

RIGHT, TOP: lithograph by Albert Chereau depicting French inventor Jacques de Vaucanson and his mechanical flute player.
RIGHT, BOTTOM: sections of the Antikythera mechanism. Recovered from a shipwreck in 1900, it has been dated to between 150-100 BC.

revolutionary, no one had quite taken the next step: to imagine and extrapolate the possibilities and pitfalls of an actual mechanical (or at least artificial) being. It would take a young woman, just nineteen years of age, to conceive and have published what would become the virtual template for a large amount of early robot fiction. Her name was Mary Wollstonecraft Shelley, and the book was *Frankenstein*, or *The Modern Prometheus*.

But before we consider Frankenstein, let's rewind a couple of centuries to around 1580 and the legend of the Golem. In the Jewish faith, a Golem is a creature fashioned from earth or clay and inscribed with the Hebrew word Emet (truth) to bring it to life. In the *Talmud*, Adam, the first man, is described as dust "kneaded into a shapeless hunk." The ancient legend of the Golem was brought to life in the 16th century story of Rabbi Loew of Prague, who—so the story goes—fashioned a Golem to protect the local Jewish community from persecution at the hands of the emperor. Ultimately, the Golem becomes so big and powerful that the emperor is forced to concede defeat and a truce is declared. The rabbi deactivates the Golem by erasing the first letter of the Emet inscription, leaving Met, the Hebrew word for death.

This apocryphal tale floated around for the best part of the next three centuries, until it finally appeared in print in a collection of

LEFT: August 10th 1965 edition of *Classics Illustrated*, featuring an adaptation of *Frankenstein* illustrated by Robert Hayward Webb.
BELOW: a scene from *The Golem*, the only existing film of the original German trilogy by Paul Wegener.

Jewish tales entitled *Galerie der Sippurim* (1847). This in turn became a novel, *Der Golem*, by Gustav Meyrink in 1915 and a trilogy of films by Paul Wegener from 1915–1920: *Der Golem, The Golem and the Dancing Girl* and *Der Golem, Wie er in die Welt* (How he Came into the World). This last film of the trilogy (a prequel to the first two) was released in the west in 1922 as *The Golem* and remains the only extant film of the three. In the film, the Golem is animated and employed as per the legend, to safeguard the Jews from the looming pogrom, but having achieved those objectives its masters involve it in increasingly more complex and emotionally charged missions and ultimately they lose control of it entirely.

The legend of the Golem maybe owes more to living statues of the ancient Greek myths than the robots of the 20th century, but themes from it can be found in *Frankenstein* and a great deal of the robot fiction that followed. Immensely strong artificial being is created, runs amok and its creator must end the threat before innocent people are hurt. That basic through-line underpins *Frankenstein*, which itself is the next evolutionary step, if you like, on the way to the classic robot archetype.

Frankenstein was subtitled *The Modern Prometheus*, a reference to the Greek Titan who–in defiance of Zeus–nurtured and expanded the horizons of mortal man, finally

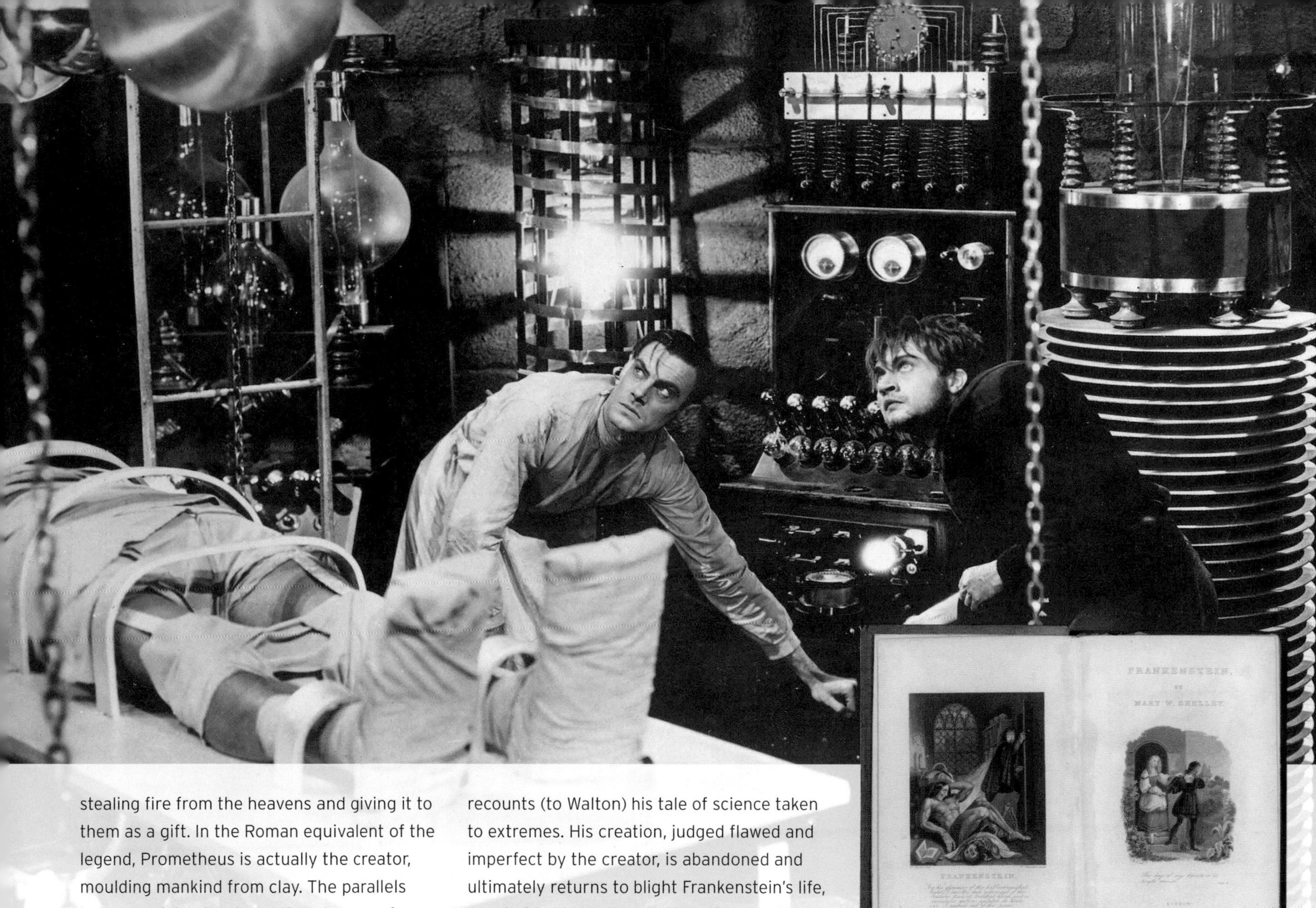

stealing fire from the heavens and giving it to them as a gift. In the Roman equivalent of the legend, Prometheus is actually the creator, moulding mankind from clay. The parallels between Victor Frankenstein—who goes far beyond the limits of science and faith to create new life—and Prometheus—who defied Zeus in order to elevate mankind—are clear. However, where Prometheus clearly cared for his 'creations', Frankenstein abandons his, precipitating murder and disaster.

In *Frankenstein*, scientist Victor Frankenstein is discovered on the Arctic ice by polar explorer Captain Robert Walton. Frankenstein recounts (to Walton) his tale of science taken to extremes. His creation, judged flawed and imperfect by the creator, is abandoned and ultimately returns to blight Frankenstein's life, murdering his brother, William (framing the Frankensteins' servant Justine in the process), his close friend Clerval and his wife Elizabeth. Finally, Frankenstein pursues the creature out into the Arctic wastelands, intent on destroying his folly. However, he dies before he can do so, aboard Walton's ship, and in the final scenes is mourned by the creature, who subsequently vows to destroy himself for his indiscretions. First published, anonymously, in 1818, this cautionary tale of science pushed

ABOVE, TOP: Colin Clive and Dwight Frye in James Whale's 1931 big screen adaptation of *Frankenstein*. A previous version, just 16 minutes long, was released in 1910.
ABOVE, BOTTOM: folio pages from an 1831 third edition of Mary Shelley's *Frankenstein*.

"Noname's" Latest and Best Stories are Published in This Library.

FRANK READE LIBRARY

Entered at the Post Office at New York, N. Y., as Second Class Matter.

No. 1. {COMPLETE.} FRANK TOUSEY, PUBLISHER, 34 & 36 NORTH MOORE STREET, NEW YORK. New York, September 24, 1892. ISSUED WEEKLY. {PRICE 5 CENTS.} VOL. I

Entered according to the Act of Congress, in the year 1892, by FRANK TOUSEY, in the office of the Librarian of Congress, at Washington, D. C.

FRANK READE, JR., AND HIS NEW STEAM MAN; OR, THE YOUNG INVENTOR'S TRIP TO THE FAR WEST.

By "NONAME."

LEFT: issue #1 of *Frank Reade Library*, published in September 1892 A total of 191 issues were published and the title was relaunched in 1902 as *Frank Reade Weekly Magazine*.

too far, too fast, without due consideration, before and after, of the consequences of such actions, is steeped in the grand gothic and romantic traditions of the time. But its themes, and those of the Golem legend, would be appropriated and embellished by the science fiction genre and its robot sub-genre.

The latter part of the 19th century saw the first real sustained clutch of robotic fiction, with works such as Edward S. Ellis' 1865 'dime' novel, *The Steam Man of the Prairies*, Luis Senarens' *Frank Reade and his Electric Man* (1885), August Villiers' *L'Eve future* (The Future Eve), 1886, and William Douglas' *The Brazen Android*. Edward S. Ellis, a prolific writer of cheap 'dime' novels, set an early trend for tales of young American inventors and their creations that get them into and out of scrapes, a torch picked up by the likes of Harry Enton, who spawned the Frank Reade series popularised by Luis Senarens in the weekly *Frank Reade Library*.

L'Eve future, which centres around a man who builds a beautiful female robot modelled on the Venus de Milo, is notable for coining the word 'android,' while Douglas' 1891 novel, published posthumously (it was originally begun in 1857), *The Brazen Android* is a retelling of the unverified legend surrounding the construction of an omniscient robotic head by the 13th century Franciscan friar Roger Bacon. Around this time also, French author Jules Verne featured a robotic or 'steam' elephant in his book *The Steam House*. As a side note, and just to stress that the idea of a robot or artificial human wasn't purely the domain of books, it's worth noting that Leo Delibes' sentimental 1870 ballet, *Coppelia*, features a life-sized, lifelike dancing doll (the titular Coppelia).

The turn of the century saw more robotic fiction, or at least fiction featuring robots, including L. Frank Baum's *Oz* books (the Tin Man and Tik-Tok) and Gustave Le Rouge and Gustave Guitton's *La Conspiration des Milliardaires* (and its Metal Men automata), but it wasn't until 1921 that robots really came of age... and got their name. Czech author and playwright Karel Capek created *R.U.R.* (Rossum's Universal Robots), finally coining the name 'robot' (based on the Czech word, *robota*, meaning forced labour). Though the robots in *R.U.R.* aren't metal and fall into the category of human robots, possibly simply due to considerations re the staging, Capek's story is the first to extrapolate on the premise of wholly artificial beings and explore both the potential benefits and—demonstrably—the downsides.

Karel Capek was born in 1890 in a small town in Northern Bohemia. Along with his brother Josef, he published numerous articles and witty, ironic prose pieces under the name 'The Capek Brothers'. In 1908, the brothers published a short story titled 'The System', which was essentially *R.U.R.* in embryo, but without the robots. In 'The System' the theme of dehumanization of the workforce in the name of enhanced productivity is explored, and in *R.U.R.* that same basic idea is rolled out to an almost absurdist degree.

In the play, the company—Rossum's Universal Robots—produces perfect artificial humans stripped of any qualities inessential to the manufacturing process, qualities such as emotion, sense of humour and creativity. They are only allowed to feel pain because it's a "built-in safeguard against damage." Capek presents first the grandiose vision, in which—while the burgeoning robot

ABOVE: Czech playwright Karel Capek, whose brother Josef first coined the word "robot".

workforce will inevitably result in mass unemployment for the more limited human beings—productivity will increase to such an extent that the price of everything from fuel to food will tumble, until they have no value at all, ultimately taking away any necessity to work and earn money. Domin, the factory owner, foresees a future when "Adam" will return to "Paradise," with no work, no cares, "master of creation."

Then, Capek, in the space between what is termed the 'comic' prologue and Act 1 of the play proper, unveils the true nature and price of this ideological Eden. Set 10 years after the prologue, Act 1 reveals that the robots—upgraded, trained for warfare and generally running the world—have risen up against the humans, organized themselves into one massive 'union' with specified aims and a manifesto of their own, and initiated a wholesale slaughter of the highly inferior mankind, which itself has grown fat, complacent and unable to defend itself. Sterility is even presented as a side effect of this mass, indulgent apathy. Only a few

RIGHT: *R.U.R.*, translated from the Czech by Paul Selver and adapted for the English stage by Nigel Playfair, was first staged in 1923 by the Reandean company at St. Martin's Theatre, London. **OPPOSITE:** a 1922 production of *R.U.R.* staged in Berlin's Theater am Kurfürstendamm, featuring stage designs by noted modernist architect Frederick Kiesler.

besieged men and women remain alive, humanity clinging on by its fingertips.

Written in 1920 and first staged in 1921, *R.U.R.* is a landmark in terms of robotic evolution. More social satire that science fiction, though an early Czech poster for the play sets the action in the year 2000, *R.U.R.* was nevertheless revolutionary in terms of its invention and outright prescience. The robots are described as having the capacity to absorb and store huge amounts of information, any small amount of which can be retrieved with a simple question. The science involved in their construction and programming is detailed and credible. And more to the point, *R.U.R.* reveals the downside of this technological evolution and the chilling cost to mankind. Echoes of *R.U.R.* can be found in the *Terminator* films, 'new' *Battlestar Galactica* and many more contemporary sci-fi opuses.

In both name and actuality, the robot had arrived, and from here on in things were only ever going to get bigger.

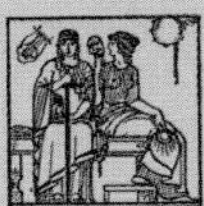

ST. MARTIN'S·THEATRE
LONDON

LESSEE B·A·MEYER· — MANAGERS REANDEAN·LTD

R. U. R.
(Rossum's Universal Robots)
A Fantastic Melodrama
By KAREL CAPEK.
Translated by PAUL SELVER. Adapted by NIGEL PLAYFAIR.

THE CHARACTERS IN THE ORDER OF THEIR APPEARANCE

Harry Domain *(General Manager of Rossum's Universal Robots)* By Mr. Basil Rathbone *(By permission of Mr. Gilbert Miller)*
Sulla *(a Robotess)* .. .. .. „ Miss Beatrix Thomson
Marius *(a Robot)* .. .. .. „ Mr. Gilbert Ritchie
Helena Glory .. .. .. „ Miss Frances Carson
Dr. Gall *(Head of the Physiological and Experimental Department of R.U.R.)* „ Mr. Charles V. France
Mr. Alquist *(Head of the Works Department of R.U.R.)* „ Mr. Brember Wills
Jacob Berman *(Chief Cashier for R.U.R.)* .. „ Mr. Malcolm Keen
Emma .. .. „ Miss Ada King
Radius *(a Robot)* .. .. .. „ Mr. Leslie Banks
Helena *(a Robotess)* .. .. .. „ Miss Olga Lindo
Primus *(a Robot)* .. .. .. „ Mr. Ian Hunter
Robots .. Messrs. Lawrence Baskcomb, Leslie Perrins, Alan Howland, Charles Cornock, Roy Leaker, Hugh Williams, George Cowley, Hugh Sinclair, Ernest Digges, David Franklin, Geoffrey Dunlop, Frederick Fanton, Cyril McLaglan, Caswell Garth.

ACT I. Domain's Room in the Offices of Rossum's Universal Robots.
Here there will be an interval of ten minutes.

ACT II. Helena's Drawing-Room. Ten years later. Morning.
After Act II. there will be an interval of five minutes only.

ACT III. The Same. Towards Sundown.
Here there will be an interval of ten minutes.

ACT IV. A Laboratory. One year later.

Place: An Island. Time: The Future.

◆ ◆ ◆

The production devised by BASIL DEAN.

The semi-permanent scenery designed by GEORGE W. HARRIS.

The imaginative costumes of the Robots made by Messrs. B. J. SIMMONS of Covent Garden, from designs by GEORGE W. HARRIS.

Miss Frances Carson's dresses by BERTHE, of Half Moon Street; her hat by ZYROT.

Chemical Apparatus and Microscope used in the Fourth Act kindly lent by Messrs. R. B. TURNER & Co., Eagle Street, W.C.

Electrical Research Apparatus by the General Electrical Company.

The Play presented by arrangement with the Directors of the Lyric Theatre, Hammersmith.

At the AMBASSADORS THEATRE — EVERY EVENING AT 8.45. MATINEES: FRIDAY AND SATURDAY AT 2.30
The REANDEAN Company in "THE LILIES OF THE FIELD"
A New Comedy by J. HASTINGS TURNER.
Miss Meggie Albanesi Miss Edna Best Mr. J. H. Roberts Miss Gertrude Kingston
The Play produced by BASIL DEAN. *The Scene- and some Costume- Designs by* GEORGE W. HARRIS.

THE PLAYBOX—Opening Shortly.
For further particulars see page 16.

Weird worlds, fantastic futures and other progressions

Robots Go Galactic!

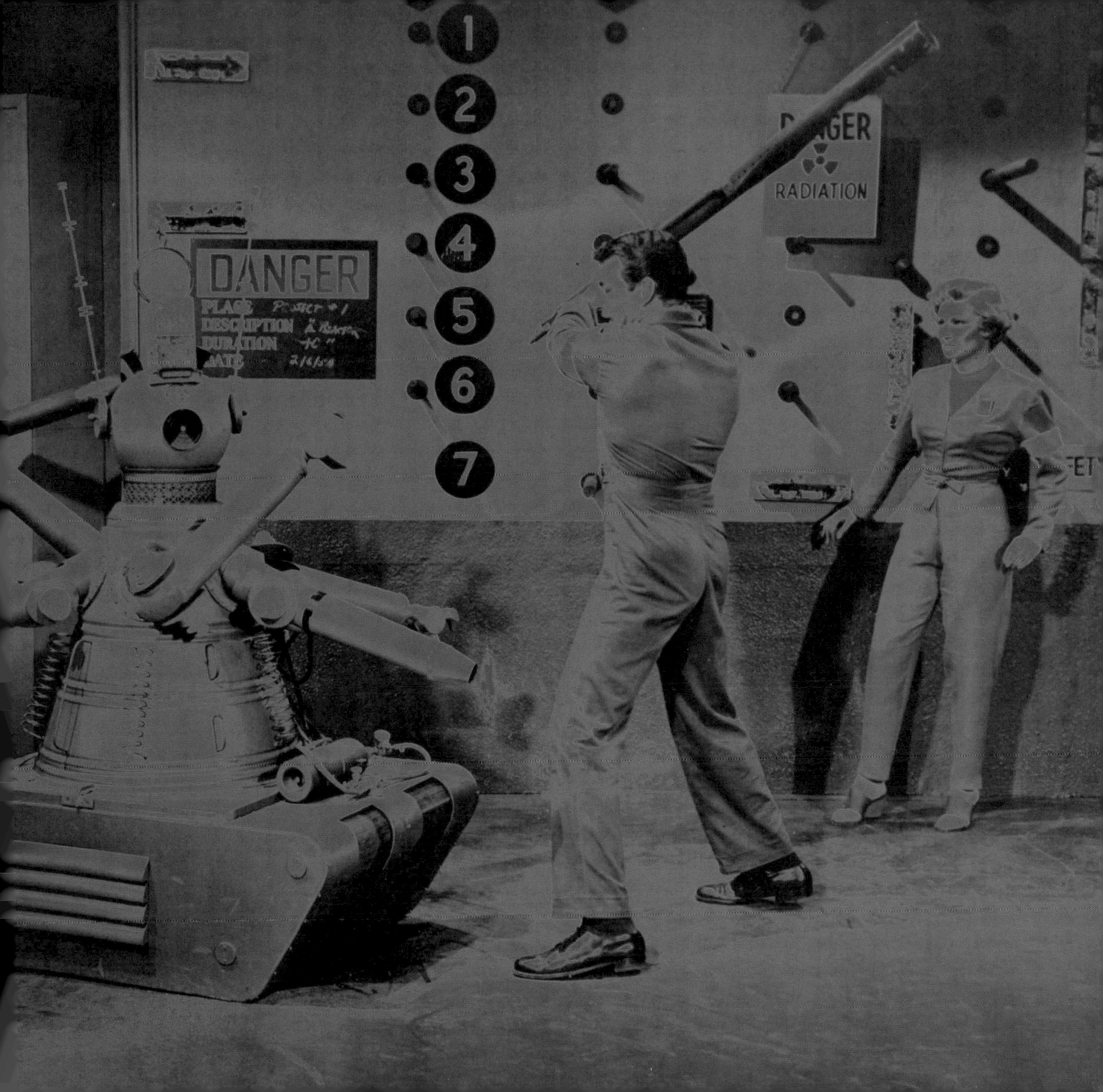

1
2
3
4
5
6
7
DANGER
PLACE
DESCRIPTION
DURATION
RADIATION

As the world moved into the 20th century, so the sheer scope of human vision and ambition turned ever further outwards, reaching for the stars.

Advancements in technology, breathlessly gathering pace every decade, meant that the scientific base for creative thought was expanding geometrically, and soon science fiction as a bona-fide genre was well and truly established.

Robots too were undergoing something of a Renaissance. Karel Capek's stage play, *R.U.R.* (Rossum's Universal Robots) had opened the door into a world of more sophisticated and science-based artificial beings and for the first time their presence had real scale and gravitas. Capek assessed both the upside and the downside, graphically showing the impact of robots—if unchecked—on the whole world. Rather than simply inventions gone awry, his robots evolved, quested and finally rebelled, turning on their creators, having surpassed them in every way. In the space of just three acts (and a prologue), ordinary, flawed flesh and blood had ceased to be relevant. It was a chilling and sobering prophecy, that one day our own science would rule us out of the equation entirely. Largely though, in terms of popular culture, it was just hugely inspirational.

Meanwhile, another technological age was dawning, that of the cinema. The advent of moving pictures and special effects (albeit basic) meant that the printed word could now be translated into cinematic visions, whole futures and other worlds realised in celluloid. A few early attempts to portray robots on the big screen came and went without overmuch of a ripple, the equivalent of—albeit remarkable—sideshow tricks. Cinematic pioneer Georges Méliès produced perhaps the first ever screen robot in his 1897 short *Gugusse et l'automaton* and other such noble follies including *The Mechanical Mary Anne* (1910) and *Sammy's Automaton* (1914) advanced the process and raised the profile, but it wasn't until 1927 that both cinematic science fiction and big screen robots really took off, courtesy of German film director Fritz Lang. His vision... was called *Metropolis*.

This silent classic explores similar themes to *R.U.R.*, with the robot as a metaphor for the dehumanization of the workforce, but in scale, scope and execution *Metropolis* is a very different beast. For its time, *Metropolis* was—and remains—a towering cinematic achievement, showcasing amazing sets and breathtaking production design. The sweeping future city of the title is a modernist masterpiece and throughout there's a sense of everything in it being much larger than life, whether in the form of towering skyscrapers or the monumental machinery at its heart. *Metropolis* is all about contrasts: the gleaming upper city and the grim, shadowy lower city; the haves and the have-nots; the surface gloss and the gritty reality. The yawning divide between 'the planners', the privileged autocrats who inhabit the high city aeries, and 'the workers', the downtrodden mass who keep the wheels turning day and night, is sharply drawn, and into this simmering, unwieldy powder keg of a world comes the ultimate expression of compact, technological perfection: a robot.

The story of *Metropolis* centres, primarily, around three individuals: Johann 'Joh' Fredersen, who runs the future city, his son Freder and Maria, the crusading woman whom Freder falls for and who takes up

BELOW: 1927 French poster for Fritz Lang's *Metropolis*, featuring striking design work by Russian artist Boris Konstantinovitch Bilinsky, commissioned by film distributor ACE (*L'Alliance Cinématographique Européenne*).

Metropolis is very much the original Dystopian future/ future city

ABOVE: colour lobby card for Fritz Lang's *Metropolis*, trimmed by some 60 minutes by Paramount for the film's US release. New title cards by playwright Channing Pollack only added to the audience's subsequent confusion.

the cause of the downtrodden workers, campaigning not for rebellion but for mediation. When Freder joins Maria's cause, Fredersen—alarmed by some propagandist literature distributed among workers, calling for industrial action—takes action of his own, contacting an old business rival and scientist, Rotwang, who is building a female robot in the image of a former lover. Fredersen convinces Rotwang to give the robot the appearance of Maria instead and imprisons the real Maria, sending the robot Maria into the under-city in her stead. However, the robot Maria whips the workers into a steamy revolutionary frenzy, inciting a full-scale revolution, and chaos ensues.

As with *R.U.R.*, the workers are presented as a mass, cogs in the vast machinery they maintain and operate. We rarely see the workers' faces as they toil in repetitive synchronised unison. The robot Maria or 'Maschinenmensch' seemingly rebels against her programming and her creator, ultimately bringing the whole city to its collective knees. Technology is again seen as the destroyer, both figuratively (of the soul) and literally (of the world). However, where *R.U.R.* was primarily a social satire, *Metropolis* is more full-blown science fiction, as concerned with the overall scope and realization of the future world in which it is set as its underlying themes. Its robot too owes more to the Grand Guignol traditions of *Frankenstein* and *Der Golem*, with hubris on the part of the creator, or at least sponsor resulting in his own downfall and the accompanying collateral damage. The overblown Gothic sets for Rotwang's laboratory were virtually a template for the countless cinematic mad scientists to come.

And *Metropolis'* influence doesn't end there. The visual impact of the city can be seen in *Blade Runner* and *Judge Dredd*'s Mega-City One and *Star Wars'* droid C-3PO, seen in embryo in the first prequel, *The Phantom Menace*, incorporates clear design elements from robot Maria. In fact, *Metropolis* is very much the original Dystopian future/future city, one that continues to be referenced and emulated even today. Sadly, *Metropolis* was barely seen in its original two and a half hour long form, and then only on its initial German release, with U.S. partner Paramount slashing nearly an hour off the running time and having American playwright Channing Pollock rewrite, rather than simply translate, the title cards. It was dismissed in some circles as fascistic on its release in Germany and subsequently—and to its lasting detriment—found favour with Nazi propaganda minister Joseph Goebbels. It also cost almost four times its original budget to make (the equivalent today of around $200 million) and so struggled both during production and after its release to recoup its investments. Latterly, though, *Metropolis* is rightly regarded as classic piece of both cinema history and science/robotic fiction.

Brigitte Helm, the seventeen year-old debutante actress who plays Maria and her robot counterpart, suffered for her art. As well as having to endure scenes where she is nearly drowned, burned at the stake and generally manhandled by the mob, *Metropolis* director Fritz Lang insisted she actually be encased in a suit of plastic-wood armour, spray-painted bronze and silver, for the scenes where the robotic Maria is seen in her full metal glory, even though the actress' actual facial likeness did not feature. The suit was hot, uncomfortable and

Metropolis director Fritz Lang insisted she actually be encased in a suit of plastic-wood armour, spray-painted bronze and silver

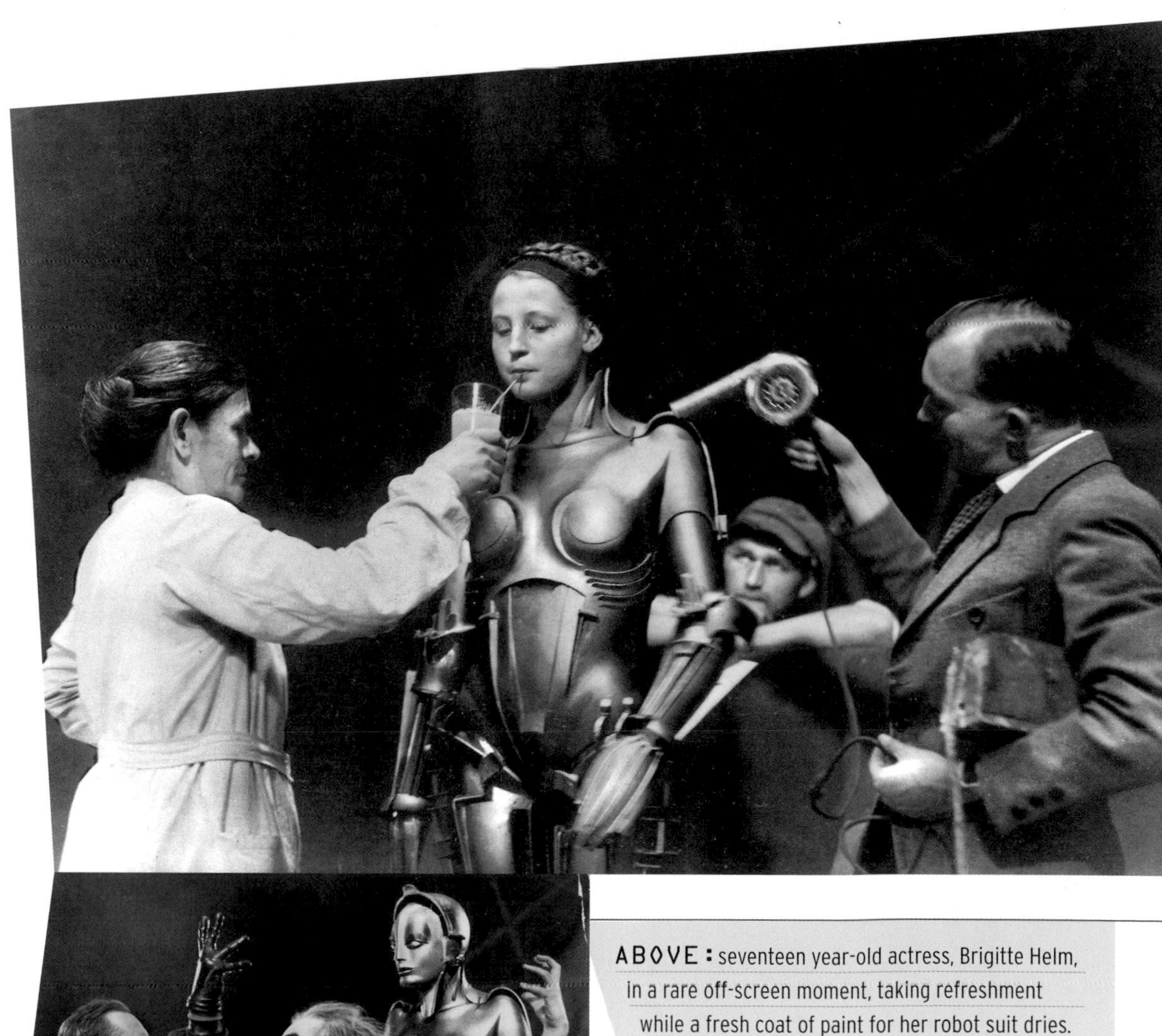

ABOVE: seventeen year-old actress, Brigitte Helm, in a rare off-screen moment, taking refreshment while a fresh coat of paint for her robot suit dries. **LEFT:** business (and love) rivals Joh Fredersen (Alfred Abel) and C. A. Rotwang (Rudolf Klein-Rogge) with their flawed creation, robot Maria.

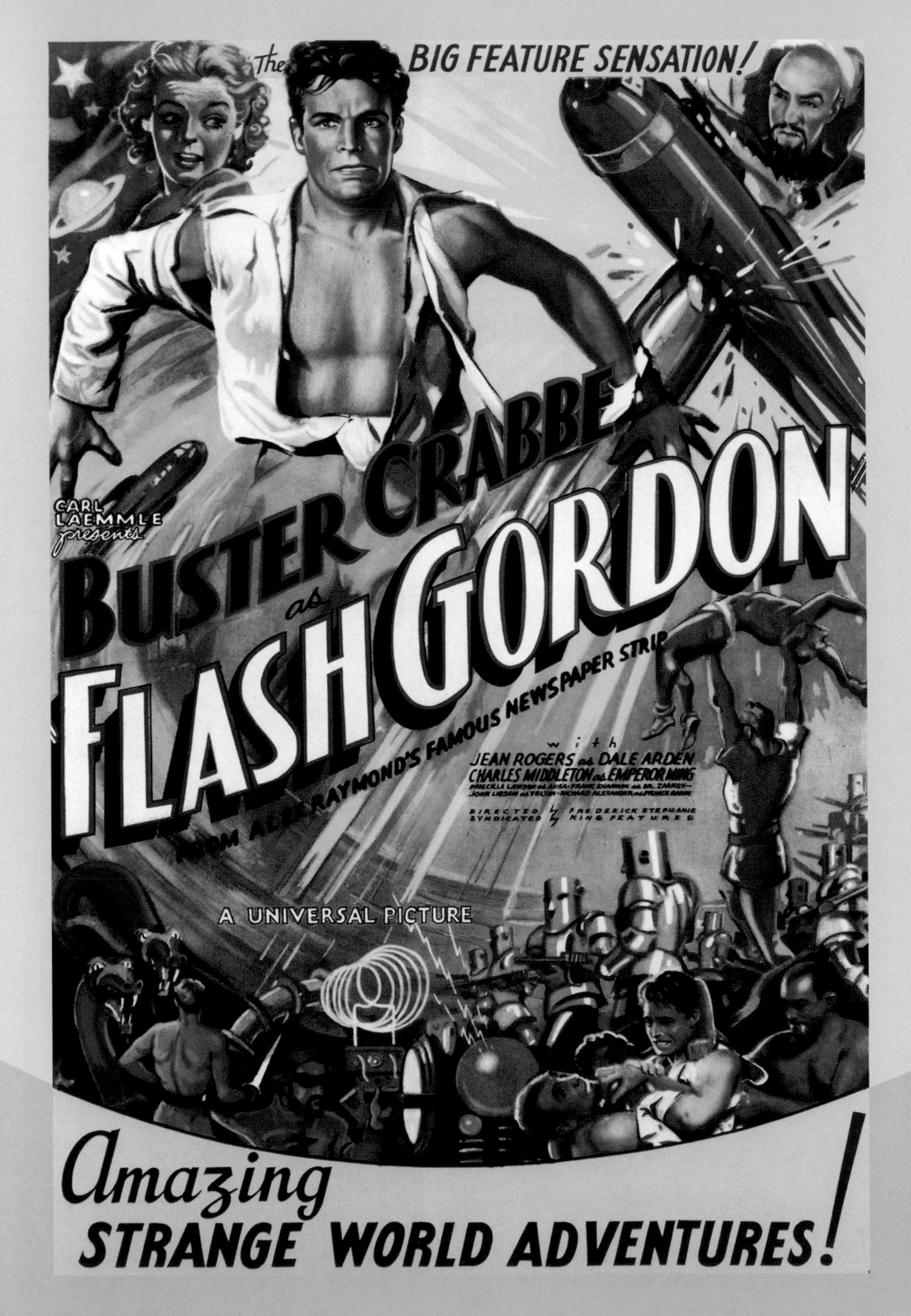

LEFT: publicity poster for the 1936 (13-episode) film serial, *Flash Gordon,* which featured the robotic Annihilants, mindless servants of Ming the Merciless. This second feature was so popular it often took top billing.

restrictive (in terms of both movement and breathing) and left Helm with numerous cuts and bruises.

The cool initial reaction to *Metropolis,* rendered virtually impenetrable by the sweeping cuts and rewritten captions, by English-language audiences meant that rather than kick-start a glut of sophisticated, thought-provoking robot movies, *Metropolis* rather did the opposite. In the 30s and 40s robot movies languished in the realm of B-movie purgatory, strangled by their own pulp fiction roots. Cast in the role of nameless foot soldiers or out-of-control alien or terrestrial menaces, robots populated such film serials as *Flash Gordon* (the Annihilants) and *Buck Rogers*, or cheapie potboilers like *The Phantom Creeps* or *The Mysterious Doctor Satan*. Only the 1949 British comedy movie, *The Perfect Woman*, in which a mad scientist creates a synthetic woman, bucks the general trend. However, other than as an example of just how sexist and politically incorrect movies of that era could be, there's really nothing to recommend it. It wasn't until 1951 that a sci-fi movie—one that at least featured a robot with a brain behind it—surfaced.

The Day the Earth Stood Still memorably introduced us to Gort, a giant robot with the power to destroy Earth should its inhabitants not heed the warnings of alien emissary Klaatu (Michael Rennie). Notable for its intelligent script and strong anti-war/anti-atomic weaponry message, *The Day the Earth Stood Still* is a prime example of movies produced in America during the height of the Cold War, when the looming threat of all-out nuclear war and mutually assured mass destruction were uppermost in hearts and minds. Klaatu's message to the people of Earth is stark and unequivocal: disarm or face the judgement of the alien races he represents. Mistaking Klaatu's 'gift' for a weapon, a soldier shoots and wounds Klaatu and Gort begins systematically decommissioning weapons and tanks on his own accord, using a kind of in-built disintegrator ray. Klaatu stops Gort before the confrontation can escalate, determined his peaceful overtures can and will be made.

Throughout the film, the robot Gort remains a kind of looming threat, rarely actually taking centre stage. Should Klaatu fail or be killed, he will proceed to systematically eradicate mankind. Gort is a weapon of mass destruction in his own right, the ultimate deterrent. Interestingly, in the short story the film is based on, 'Farewell to the Master' by Harry Bates (originally published in the October 1940 issue of *Astounding Science Fiction*), Gort (or Gnut in the story) turns out to be, in a twist ending, the master and Klaatu the servant. In both short story and film, the robot is—refreshingly—not simply the mindless thrall of some human or alien madman or conqueror. Instead he has the kind of stoic, unswerving purpose that would come to characterise many later robots, neither good nor bad, simply obeying a higher directive. And what's more, Gort is wholly extraterrestrial, not a creation of some deranged or misguided earthbound science.

BELOW: the imposing presence of Gort (Lock Martin) reinforces the message of Klaatu (Michael Rennie) in director Robert Wise's superlative Cold War era movie *The Day the Earth Stood Still.*

Things were finally moving in the right direction, but Gort—for all his considerable presence—is still hardly a character in his own right. For that, audiences would have to wait until 1956 and the screen debut of the first robot superstar—Robby. In the interim, we were cast once more into the B-movie backlot with the likes of Ro-Man, Tobor the Great and Gog/Magog, none of whom did much to advance the cause. Ro-Man appeared in the "so bad it's good" movie *Robot Monster*, and—due to budgetary constraints—amounted to a man in a gorilla costume wearing a diving helmet. That said Ro-Man, largely due to his kitsch value, stuck somewhat in popular culture, appearing subsequently in Woody Allen's *Stardust Memories*, featuring in music videos for both Cyndi Lauper's 'Girl's Just Wanna Have Fun' and The Cars' 'You Might Think' and cropping up on *Mystery Science Theatre 3000*, in which the robotic hosts are 'forced' to watch terrible movies.

Tobor the Great, from *Tobor, The Great*, was at least more robotic in appearance, but still had that 'cobbled together from any spare tat hanging around the costume department' look. Whereas Ro-Man was a conquering alien (or aliens) from outer space, Tobor was a homegrown affair, designed to survive the 'rigours' of space travel, who ends up battling Commies (boo-hiss). *Gog* featured two robots, the Biblically named Gog and Magog, and a rogue A.I., N.O.V.A.C. (Nuclear Operative Variable Automatic Computer), and is by far the best of the bunch. Another offspring of Cold War paranoia, Gog involves a secret government facility in New Mexico

ABOVE, LEFT: it's a man in a monkey suit and a diving helmet! No, it's Ro-Man, from no-budget creature feature *Robot Monster.* **ABOVE, RIGHT:** Tobor the Great, a crash test dummy for the space age, gets to grips with the red menace in *Tobor the Great.* **RIGHT:** Gog (or Magog, it's hard to tell, and barely matters) attacks in *Gog,* which also featured a rogue A.I. in the shape of N.O.V.A.C. (Nuclear Operative Variable Automatic Computer).

1
2
3
4
5
6
7
DANGER
PLACE
DESCRIPTION
DURATION
RADIATION

and an attempt to sabotage the space station being built there. The multi-armed Gog and Magog, at the behest of N.O.V.A.C. (who, in turn, is being controlled by an external signal), attempt to initiate meltdown in the facility's nuclear reactor, but are thwarted by security agent, David Shepherd (Richard Egan).

Making his debut in *Forbidden Planet*, the somewhat stagey but still well regarded sci-fi retread of Shakespeare's *The Tempest*, Robby was notable for two reasons. In the first instance, he was, at the time, the closest cinematic equivalent to the kind of servile, logical but sometimes contrary and 'difficult' robots author Isaac Asimov was busily writing about in the likes of *Astounding Science Fiction* and *Amazing Stories*. In fact, Robby operates under his own somewhat more abridged version of Asimov's three laws of robotics (see chapter 3), and will not under any circumstances allow harm to come his human charges. And secondly, Robby—almost like an actor—would emerge from *Forbidden Planet* into a career (on big and small screen) in his own right.

In *Forbidden Planet*, Robby is the all-round domestic servant, security consultant and walking computer of Doctor Morbius (Walter Pidgeon), who—along with his daughter, Altaira—is the last survivor of a colony expedition to the planet Altair, the other colonists having died in mysterious circumstances. Robby cooks, cleans, gathers and generally gets all the best lines, in the process out-acting a wooden-faced Leslie Neilson as Commander John J. Adams, who arrives on Altair as part of a 'rescue' mission. Morbuis, though, doesn't want rescuing, and warns Adams he can't guarantee the safety of his crew if he lands. However, Adams does so, and one by one they fall victim to an invisible monster, which turns out to be the expression of Morbius' unconscious id, amplified by alien machinery within the planet.

When Robby is called upon to defend the colonists and Adams' crew from the monster, his systems overload and he goes offline. Why? Because, he is—in effect—turning on his own creator, something strictly forbidden by and contrary to his programming. Again, this is where the influence of Asimov's stories is most telling. Robby ultimately survives the movie and is whisked away from Altair aboard Adams' ship. Away also, we assume, from the daily drudgery of having to cook and clean and make dresses for Altaira, Robby—for all the progress in terms of his character—essentially a kind of year 2022 housewife. We all hope for a better and more fulfilling life for him on Earth.

In film production terms, Robby was designed and built by engineer Bob Kinoshita, who also provided the models for the C57D flying saucer

ABOVE: in a rare publicity shot for *Forbidden Planet*, poor Robby tries hard not to look embarrassed as he helps model yet another of Altaira's (Anne Francis) frocks!
OPPOSITE: bound by laws similar to Asimov's Three Laws, Robby can let no harm befall his creator, Doctor Morbius (Walter Pigeon). Sadly, the same doesn't apply to John J. Adams (Leslie Neilson) and crew!

in which Adams and his crew arrive on Altair. Built of metal and weighing in at 100lbs and carrying some 2,600 feet of electrical wiring (to power the spinning, sparking gadgets in Robby's dome helmet), Robby cost a staggering $125,000, which perhaps explains his re-use, a year later, in *The Invisible Boy*. Operated by Frankie Darrow, who sat inside the robot, manipulating its limbs, and voiced by actor Marvin Miller, Robby endeared himself to millions and surely influenced George Lucas when it came to protocol droid C-3PO in *Star Wars*. Robby, when introduced to Adams, declares (in case they don't speak English) he is at their disposal "with 187 other languages along with their various dialects and sub-tongues."

Sadly, whatever thought went into Robby's debut vehicle, very little went into *The Invisible Boy,* a strange and disjointed matinee timeslot filler. The plot, such as it is, involves a rogue computer or artificial intelligence, one of the very first, an oh-so 50s oversized conglomeration of tubes, lights and reel-to-reel tape drives and, strangely, a dome-like 'head' that looks a bit like Robby's head. Maybe the idea is it's Robby's dad! Anyway, to cut a dull story short, the computer conspires—via its inventor's son, Timmie—to have itself relocated to some planet, from where it can plan to eradicate all organic life on Earth without fear of counterattack.

OPPOSITE: progressive parenting. Robby tries to instil some values in precocious youngster Timmie (Richard Eyer) in *The Invisible Boy*. Fat chance! **LEFT:** na-nu, na-nu! Among Robby's many guest-star roles was this appearance in *Mork & Mindy*, alongside Robin Williams in a 1979 episode entitled 'Doctor Morkenstein'.

En route to the realization of its master plan (bwa-ha-ha!), the computer grants Timmie super-intelligence so he can rebuild Robby, who's in bits, and the power to turn invisible, largely it seems so they could call the film *The Invisible Boy*.

Little if any explanation is offered of why Robby is in bits or why and how he came to be on Earth, in the 1950s. There's the briefest suggestion of time travel, and that Robby was brought back from the future, but it's glossed over in the same way that a handy 'invisibility machine' is just there, sitting around in the basement, and Robbie's parents seem to accept his invisibility as any other childhood prank. At one point, Timmie's mother mistakes Robby for a door-to-door salesman! At least Robbie proves a more progressive parent, preventing Timmie's actual father from meting out a—let's face it, well deserved—spanking. Anyway, enough of *The Invisible Boy*! Least said, soonest mended.

Thankfully, *The Invisible Boy* didn't blight Robbie's career. Apart from becoming a hugely popular toy, reissued many times since, Robby went on to guest star in numerous TV shows. He appeared in several episodes of *The Twilight Zone*, among them 'The Brain Center at Whipples,' in which a plant manager, Wallace V. Whipple, decides to replace his workers with robots. Hmm. The twist in the tale is that Whipple is also replaced. By Robby! Robby also appeared in the episodes 'One for the Angels', though just as a toy, and 'Uncle Simon', as the invention of a murdered man that comes back—complete with the dead man's memories—to 'haunt' the wife that killed him. Robby cropped up elsewhere too, in *The Addams Family* (as a household assistant to Lurch), *Gilligan's Island* (as a prototype government robot), *Columbo* (as a computerized invention) and *Mork & Mindy* (as a relic in a museum given emotions), as well

as later movies such as *Hollywood Boulevard*, *Gremlins*, *Earth Girls are Easy* and *Looney Tunes, Back in Action*.

Sadly, Robby's greatest amount of screen time would be uncredited. The TV series, *Lost in Space*, which ran from 1965 to 1968, famously featured Robot B-9, a blatant Robby look-alike. Hardly surprising when you consider that Robot B-9's designer was none other than Robert Kinoshita, the man who'd designed Robby for *Forbidden Planet*. To be fair, B-9's head is significantly different, a sort of raised, flattened bowl rather than a dome. Famous for continually spouting, "Danger, Will Robinson, danger!" and being routinely subverted by Doctor Zachary Smith (Jonathan Harris), B-9 made just one reappearance, in the 1998 *Lost in Space* movie before forever fading into obscurity. So I guess Robby had the last laugh after all.

Post-Robby, things reverted pretty much to form. The 1958 film, *The Colossus of New York*, neatly sums up the 50s attitude to science and technology. On the one hand, it's a good thing. On the other—watch out! The good is represented by Doctor Jeremy Spensser, a scientist striving to feed the world via new strains of edible plant, and the bad is represented by Doctor Jeremy Spensser... after, that is, his crackpot, cardboard cut-

OPPOSITE: a scene from Eugene Lourié's *The Colossus of New York*, with Ed Wolff as the robotic version of (dead) scientist Jeremy Spensser, complete with death-ray eyes. No robot should be without them!
RIGHT: Robot B-9 from *Lost in Space,* alongside the conniving Doctor Zachary Smith (Jonathan Harris). Sadly, robot B-9's operator, Bob May, passed away in early 2009, aged just 69.

out mad scientist father transfers his brain to a robot body following a fatal car accident. It all ends badly, of course, not least because Spensser senior thoughtfully kits out his son's robot body with a death ray, E.S.P. and super-strength!

From there, it was pretty much all downhill until the late 60s/early 70s, when invention (in the good sense of the word) started to once more creep into the concept and packaging of cinematic robots and A.I.s, with the likes of *2001: A Space Odyssey*, *Colossus: The Forbin Project* and *Silent Running*. But, for all the dross, the 50s did at least put robots on the map, crafting an enduring icon in Robby and pushing the whole concept of artificial beings into space and beyond. Early robots are a reflection of our own faltering steps into the technological twentieth century, at once appealing and terrifying, boon and threat all in one shiny metal package.

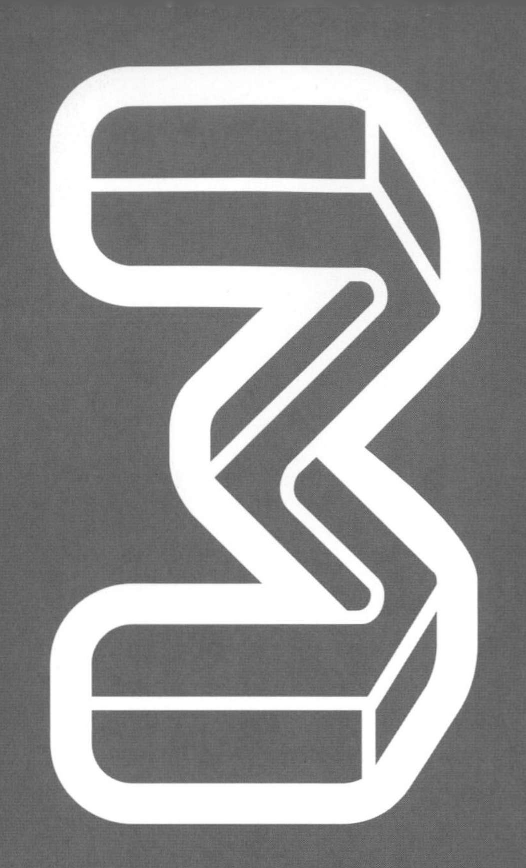

Asimov's Robots

Pulp fiction,
the Three Laws
and beyond...

LEFT: cover to the 3rd October 1936 issue of *Argosy*, one of the original 'pulps'.

It was really a very simple step, but it took a lot of years before someone took it. The standard robot yarn, prevalent in the 1930s boom of pulp science fiction magazines, went something like this: mad scientist, a la Victor Frankenstein, creates a robotic or artificial being, heedless of his own misgivings or the warnings of others.

Through the creator's own miscalculation or some other external influence, the robot malfunctions, control is lost, and the scientist and often many of his nearest and dearest pay the ultimate price. The message, couched either in fear or ignorance (or both), was almost textbook: don't mess with what should be left in the hands of God. But one up-and-coming author grew tired of this by-the-numbers approach to the robot story, and asked of himself a simple question: why, as in other machines, were there no in-built safeguards? And so, though not all at once, the three laws of robotics were born. The author was Isaac Asimov.

Pulp magazines, named after the cheap paper stock on which they were printed, had already been a publishing staple for a good long while before Asimov first started contributing in the late 1930s. It was, back then, often the way new writers broke in, cutting their teeth on short, serialized or interlocking stories and, when a significant amount of material was available, publishing them in a collected book form. No one's quite been able to identify exactly when pulps began. It was somewhere in and around the 1880/1890s, and the most likely candidate, largely due to a shift in format from newspaper to magazine (7" x 10") size that set a template for the pulps that followed, is *The Argosy*, which began life in 1882 as *The Golden Argosy* and mixed fiction with general information. By 1894, *The Argosy* was monthly and all-fiction, and aimed firmly at adults rather than children. *The Argosy*, which—along the way—absorbed other titles such as *All-Story Weekly*, featured the work of Edgar Rice Burroughs, Zane Grey, Max Brand and many other genre luminaries.

Pulps of all hues and showcasing a wide range of styles and genres proliferated until in 1923 a magazine called *Weird Tales* was launched, featuring solely horror fiction, and was followed in 1926 by *Amazing Stories*, whose stated remit to publish only "scientification" stories, and in 1930 by *Astounding Stories* (retitled *Astounding Science Fiction* in 1938). *Astounding's* editor, John W. Campbell, insisted on "hard science" as the basis for the stories it carried, and it was in this rigorous environment that Asimov and other fine authors such as Robert Heinlein and Frank Herbert thrived. Pulps also thrived, until the advent of the Second World War and the accompanying paper shortage all but wiped them out. Some did continue in a smaller digest size, and *Astounding Science Fiction* continues to this day as *Analog Science Fact and Fiction*.

Isaac Asimov was born in Petrovichi in Russia in 1920, his family moving to the United States in 1923. He sold his first short story, 'Marooned Off Vesta,' to *Amazing Stories* in 1938 and it was published in early 1939. Asimov's first robot story, 'Strange Playfellow', was published in the September 1940 issue of *Super Science Stories*, but had actually been written, as 'Robbie', a year before. Originally submitted to *Astounding*, it was rejected by Campbell, forcing Asimov to hawk his tale elsewhere. 'Strange Playfellow', or 'Robbie' as it became again on subsequent publication, introduced the then revolutionary concept that all robots were created with a strict in-built, inviolate conditioning, "laws" that meant they could not by direct action or inaction harm a human being. This safety

TOP : author Isaac Asimov, co-creator of the three laws of robotics.
ABOVE : January 1939 issue of *Amazing Stories*, featuring Eando Binder's "I, Robot", a title later appropriated for the collected robot stories of Isaac Asimov.
RIGHT : *Astounding Science Fiction*, as edited by John W. Campbell, helped authors such as Asimov, Robert Heinlein and Frank Herbert find an outlet.

with the lives of humans in a distinctly non-destructive manner, sets a new tone for the genre. In one bold stroke, the shambling, monster-of-the-week trappings were stripped away, paving the way for a brave new world of robot storytelling. In place of fear was optimism. In place of destruction was creation. The robot story had come of age.

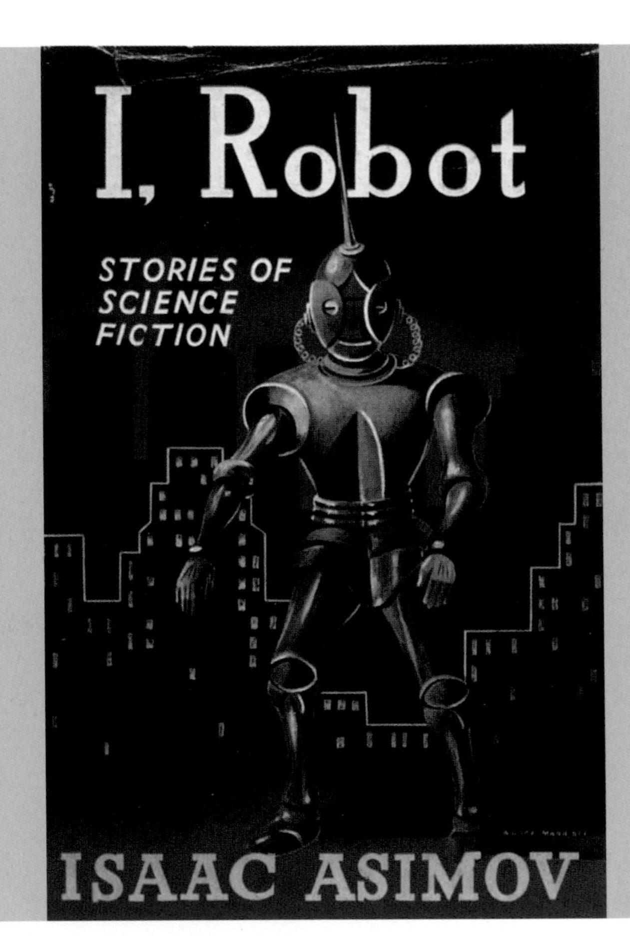

ABOVE: an early edition of Asimov's collected robot fiction. The nine stories were strung together with a new bridging sequence featuring recurring character robo-psychologist Susan Calvin.

that, in Asimov's stories at least, a robot simply couldn't turn against its creator or any other human being. Doing so would be in direct contravention of its most basic programming.

In 'Robbie,' the robot in question is about as far from an out-of-control pulp menace as one could imagine. The titular Robbie is a childminder, the dedicated, tolerant and protective companion of precocious young Gloria. However, all is not well as far as Gloria's mother, Grace, is concerned. While robots in Asimov's 'positronic' future are commonplace and, if not accepted, tolerated, it's apparent that much of the general population still fears them, or regards them with suspicion. In 'Robbie,' a groundswell of public backlash against robots is building, and Grace reacts to that, echoing concerns over her daughter's wellbeing. Her long-suffering husband, George, responds that Robbie simply cannot hurt Gloria, as he and all other robots are slaves to The First Law: 'A robot may not injure a human being, or, through inaction, allow a human being to come to harm.' Ultimately, George proves this by engineering an incident in which Robbie risks certain destruction rather than allow Gloria to come to any harm and so Grace allows Robbie to stay on.

This knowingly gentle and slight story, in which functional, everyday robots interweave

With his next few robot stories, Asimov began to flesh out the positronic world he'd hinted at in 'Robbie', introducing recurring human characters such as trouble-shooters Powell and Donovan and robo-psychologist Dr. Susan Calvin. He also introduced the world to U.S. Robots and Mechanical Men, Inc., the organization responsible for the proliferation of (largely industrial or off-world) robots. Asimov's second robot story, 'Reason', as well as introducing U.S. Robots field agents (Greg) Powell and (Mike) Donovan, brought in Asimov's trademark subversion of his own Laws. Having established that robots, through direct or indirect action, could never harm a human being, Asimov set out to present dilemmas that continually challenged that First Law. In 'Reason', Powell and Donovan are confronted with a robot unable to believe it was created by (inferior) human beings, such is its perceived sophistication. The robot concludes that it was created by a "higher power" and therefore chooses not to obey Powell and Donovan, unwittingly placing the whole of planet Earth and humankind in dire peril. 'Reason', with its religious overtones, made it through the stringent John W. Campbell filter and into the April 1941 issue of *Astounding*, thereby beginning a long and fruitful collaboration.

In the first of the Susan Calvin stories 'Liar' (though Calvin features very briefly as a teenage student in 'Robbie', beginning a chronology of events in the positronic universe that would be explored further in the first collection, *I, Robot,* a robot with the ability to read minds flirts with its programming by lying to Calvin and her colleagues, but—we learn—it is only doing so to prevent *emotional* damage to her and the others, and thereby stays true to the First Law. 'Liar' was published in the May 1941 issue of *Astounding*. Asimov's fourth robot story, 'Source of Power', was published not in *Astounding*—as Asimov himself deemed it not as strong as his other work and therefore didn't even bother submitting it to Campbell—but instead in *Amazing Stories* (February 1942) as 'Robot AL-76 Goes Astray', but it was during this period that Asimov and Campbell's many discussions led to the establishment of the three laws of robotics, which would feature in the next (Powell and Donovan) robot story, 'Runaround' (*Astounding Science Fiction*, March 1942).

These laws (see right) would come to underpin all of Asimov's subsequent robot stories, including 'Catch That Rabbit', 'Escape' and 'Evidence'. There's a popular misconception, reinforced somewhat by the recent Will Smith film of the same name, that in 1950 Asimov wrote a book called *I, Robot*. He didn't. *I, Robot* is a collection of nine of Asimov's short stories, with a new framing sequence in which Susan Calvin is interviewed prior to her retirement from U.S. Robots in 2057. The framing sequence serves to date the other stories and expand upon world events at large. The title *I, Robot* was not even of Asimov's choosing. The collection was originally to be called *Mind and Iron*, but the publishers—Gnome

ISAAC ASIMOV
THE THREE LAWS OF ROBOTICS:

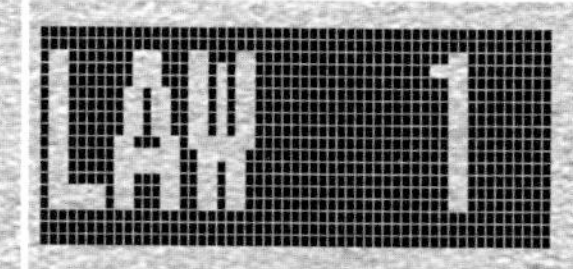

A robot may not injure a human being or, through inaction, allow a human being to come to harm.

A robot must obey orders given it by human beings except where such orders would conflict with the First Law.

LAW 3

A robot must protect its own existence as long as such protection does not conflict with the First or Second Law.

LEFT: although it bore little relation to Asimov's work, the 2004 film, *I, Robot*, starring Will Smith and directed by Alex Proyas, did incorporate some elements of and characters from his robot stories, presented onscreen through cutting edge CGI.

Books—changed it to *I, Robot*, which itself was the name of another short story, by Eando Binder (a pseudonym for Earl and Otto Binder), published in the January 1939 issue of *Amazing Stories*.

Strangely, given the range, quality and general mould-breaking nature of Asimov's robot stories, big and small screen adaptations of his work have been relatively few and far between. The Asimov back catalogue of short stories was first optioned in 1967, and many renewals later, in 1979, author Harlan Ellison was hired to write a screenplay. Asimov was apparently pleased with the result, which retained the integrity of the original stories but now featured Susan Calvin in a much more prominent role, even going as far as to insert her into stories in which she didn't originally feature at all. The screenplay was constructed around a series of flashbacks, with the emphasis more on character development than action. However, due to budgetary considerations and the available special effects technology at the time, the screenplay was deemed un-filmable. The 2004 blockbuster *I, Robot*, starring Will Smith claims to be "suggested" by Asimov's stories, but really only incorporates a few character names—Susan Calvin, Dr. Alfred Lanning and Sonny the robot—and the three laws of robotics.

Otherwise, *Out of the Unknown*, a BBC TV series that ran from 1965 to 1966 and was briefly reprised in 1969, adapted three Asimov robot stories, 'Liar', 'Satisfaction Guaranteed' and 'Reason' (re-named 'The Prophet'),

U.S. Network A.B.C. adapted 'Little Robot Lost' for their similarly titled *Out of This World* series in 1962, and the Asimov book, *Caves of Steel* (one of a series of robot/detective novels that includes *The Robots of Dawn* and *The Naked Sun*) was adapted by Terry Nation in 1964, starring Peter Cushing as Elijah Bailey. Asimov's later robot short story, 'The Bicentennial Man', was adapted as a movie, *Bicentennial Man*, starring Robin Williams. In his foreword to 'The Bicentennial Man', in *The Complete Robot* (a collection of all the Asimov robot short stories), Asimov describes the story as his "favourite" and "the best" he ever wrote. Sadly, we'll never know if Asimov would have approved of the Robin Williams-starring celluloid adaptation as he died, aged 72, in 1992.

Asimov's true literary legacy is far less about celluloid immortality and more about the grounding of the robot story in hard science and non-hysterical storytelling. Shorn of their mad scientist/monster-movie baggage, robots emerged into the latter half of the 20th century as objects worthy of deeper consideration, a technological marvel forging a bold path into the 21st century and beyond. Asimov made us think about the practical realities of a world struggling to embrace an evolution, one that—while inevitably flawed and unpredictable—offered hope and promised wonder. And in doing so he inspired other writers, other creators, to challenge our knee-jerk idea of what a robot is and could be.

Asimov, of course, was not alone in advancing the public perception of robots. Author Jack Williamson, in his novel *The Humanoids*, introduced an interesting twist on Asimov's benevolent, law-abiding if sometimes prickly robots. In Williamson's story, the robots of the planet Wing IV are utterly benevolent, too benevolent in fact. Their Prime Directive—"To serve and obey and guard men from harm"—is interpreted in such a literal fashion that they set out to systematically stifle human endeavour, spirit and creativity, in their own minds saving man from himself. The novel was preceded by the novelette, *With Folded Hands*, and followed much later (in 1980) by *The Humanoid Touch*. It was filmed as *The Creation of the Humanoids*.

Author Philip K. (Kindred) Dick also made a number of telling forays into the world of robots. In 'The Last of the Masters' we meet Bors, a 200-year-old Government Integration Robot, and the last of his kind. A global anarchist revolution has long since toppled all previous governmental structures and society—in the years since—has stagnated. Bors heads one last, hidden pocket state of society, as was, but now—through lack of replacement parts and technological know-how—his days are numbered. Three members of The Anarchist League attempt to find Bors' hidden enclave and destroy it, but in doing so push the world to brink of another catastrophic conflict. When Bors finally falls, the last remnants of the old society die with him and those who sought his downfall ultimately lament the passing of an epoch.

Though 'The Last of the Masters' seems more concerned with weighing its theme of anarchy against the state, the robot—Bors—serves as a metaphor for a sterile, technologically stunted and dead-ended future, one that ultimately offers little hope for humankind. Without the forward momentum of advanced sciences, such as robotics, society stagnates and, ultimately, perishes, is what Dick appears to be saying. A far cry from earlier literary themes, ones that sought to rein in scientific advancement with dire warnings of subsequent prices to pay.

Much better known is Dick's 1968 novel, *Do Androids Dream of Electric Sheep?*, largely due to its later (loose) adaptation into the movie *Blade Runner*. Though certain themes do feature in both, of the two the book is the more rich and multi-layered entity, in which the androids (replicants in the movie) are not simply second class beings consigned to off-world colonies, but a poor third class,

OPPOSITE: two screen adaptations of Asimov's work. Top: Robin Williams in *Bicentennial Man*, a loose adaptation of the short story of the same name. Bottom: Peter Cushing as Elijah Bailey in the BBC's adaptation of the novel *Caves of Steel*. **ABOVE:** poster for *The Creation of the Humanoids*, based on Jack Williamson's novel *The Humanoids*.

A chilling, bold, mesmerizing, futuristic detective thriller
HARRISON FORD IS
BLADE RUNNER™ AA
Panther Science Fiction
Philip K. Dick
Hugo Award-winning author
Do Androids Dream of Electric Sheep?
Y PERENCHIO AND BUD YORKIN PRESENT A MICHAEL DEELEY-RIDLEY SCOTT PRODUCTION
A FILM BY RIDLEY SCOTT
STARRING HARRISON FORD
IN BLADE RUNNER™ WITH RUTGER HAUER SEAN YOUNG
EDWARD JAMES OLMOS SCREENPLAY BY HAMPTON FANCHER AND DAVID PEOPLES
XECUTIVE PRODUCERS BRIAN KELLY AND HAMPTON FANCHER VISUAL EFFECTS BY DOUGLAS TRUMBULL
ORIGINAL MUSIC COMPOSED BY VANGELIS PRODUCED BY MICHAEL DEELEY DIRECTED BY RIDLEY SCOTT
RELEASE IN ASSOCIATION WITH SIR RUN RUN SHAW THRU WARNER BROS A WARNER COMMUNICATIONS COMPANY DISTRIBUTED BY COLUMBIA-EMI-WARNER
PANAVISION® TECHNICOLOR®
DOLBY STEREO™ IN SELECTED CINEMAS ONLY
© 1982 The Ladd Company. All Rights Reserved.
WIN
1983 Volkswagen P

well below animals which—on the radiation ravaged post-World War Terminus Earth in which the story is set—are status symbols of the highest order. Rick Deckard, a bounty-hunter for the San Francisco Police Department, keeps an electric sheep, his real sheep having previously died. He and his wife Iran keep up the deception, even going so far as to routinely feed the sheep, to improve their social standing. When Deckard subsequently buys a live goat, Rachel Rosen—a human in the employ of the Rosen Corporation, who make the advanced Nexus-6 androids Deckard is hunting in the novel—comments that in all likelihood he cares more for the goat than his wife.

In *Do Androids Dream of Electric Sheep?*, the themes of stagnation and decay, touched upon in 'The Last of the Masters', are even more pronounced. Religion, or what passes for it in the book, is already dead, surpassed by a group, Internet-style empathy box, allowing the users to tap into the 'death and rebirth' of quasi-deity Wilbur Mercer, who himself is maybe simply a repeating computer program, employing the talents of a former alcoholic actor, and the populace en masse resorts to a Penfield Mood Organ to feel anything much at all, selecting and scheduling their emotional highs and lows. The men wear lead codpieces in a perhaps vain effort to protect their ability to foster future generations. The only hope appears to lie in the increasingly sophisticated androids, but even they are limited by a four-year lifespan due to an inability to replace deteriorating cells. Once again, humankind has ground to a shuddering halt, and this time all its science won't save it.

Blade Runner, the film, focuses much more on the plight of the Nexus-6 androids, all of whom are questing to prolong their own, brief lives (a possibility that is never floated in the book). Both book and film flirt with the issue of identity. In the book, characters—including Deckard himself—have cause to wonder if they too are androids (though the book seems to pretty much rule out Deckard being one, whereas the film is more ambiguous). The suggestion in both book and film seems to be that we may ultimately reach a point where the differences between human and artificial being will become unnoticeable, perhaps even immaterial, and the defining factor is our own prejudices, our need to create a persecution hierarchy based on social factors.

The literary lineage of robots extends through many more fine authors, among them Ray Bradbury ('The Electric Grandmother' from *I Sing the Body Electric*) and Thomas Pynchon (*V*). *The Iron Man: A Children's Story in Five Nights* was written by poet Ted Hughes to

OPPOSITE: original film poster for Ridley Scott's hugely influential *Blade Runner*, and the book that inspired it, Philip K. Dick's *Do Androids Dream of Electric Sheep?*
BELOW: Harrison Ford as Rick Deckard in *Blade Runner*. Is he or isn't he? The debate as to Deckard's status as a replicant himself rages on.

comfort his children in the wake of their mother, Sylvia Plath's suicide, and centres around a small boy and a giant extraterrestrial robot who strike up an unlikely friendship and defend Earth from a creature known as The Star Spirit. *The Iron Man* was filmed, albeit in a re-imagined form, as *The Iron Giant* by Brad Bird (*The Incredibles*). *The Iron Woman*, a sequel to *The Iron Man* also by Hughes, was published in 1993. Brian Aldiss' short, sweet story, *Super-Toys Last All Summer Long*, about a couple whose young boy is actually a robot, due to a government embargo on natural conception, also made it as far as the big screen, but its journey was long and convoluted. Originally slated to be Stanley Kubrick's next big science fiction opus (following *2001: A Space Odyssey*), *Super-Toys Last All Summer Long* was conceived and re-conceived (at one point by Arthur C. Clarke) until finally stalling altogether. After Kubrick's death, the project was re-energized by Steven Spielberg and released as *A.I.: Artificial Intelligence*.

One of the most popular and enduring literary robots to emerge post-Asimov was Marvin the Paranoid Android, a character from Douglas Adams' radio play, novels, TV series and film, *The Hitchhiker's Guide to the Galaxy*. Marvin, a depressed and bored super-computer who woefully and continually laments the fact that no task will ever occupy more than a fragment of his vast intellect, droned his way morosely into the hearts of many. One way or another robots had come a long way since the dark old pulp days of their literary genesis, but I confess a slight yearning, every now and then, for a good old-fashioned malfunctioning, marauding mechanical monster. I mean, really, what's not to like?

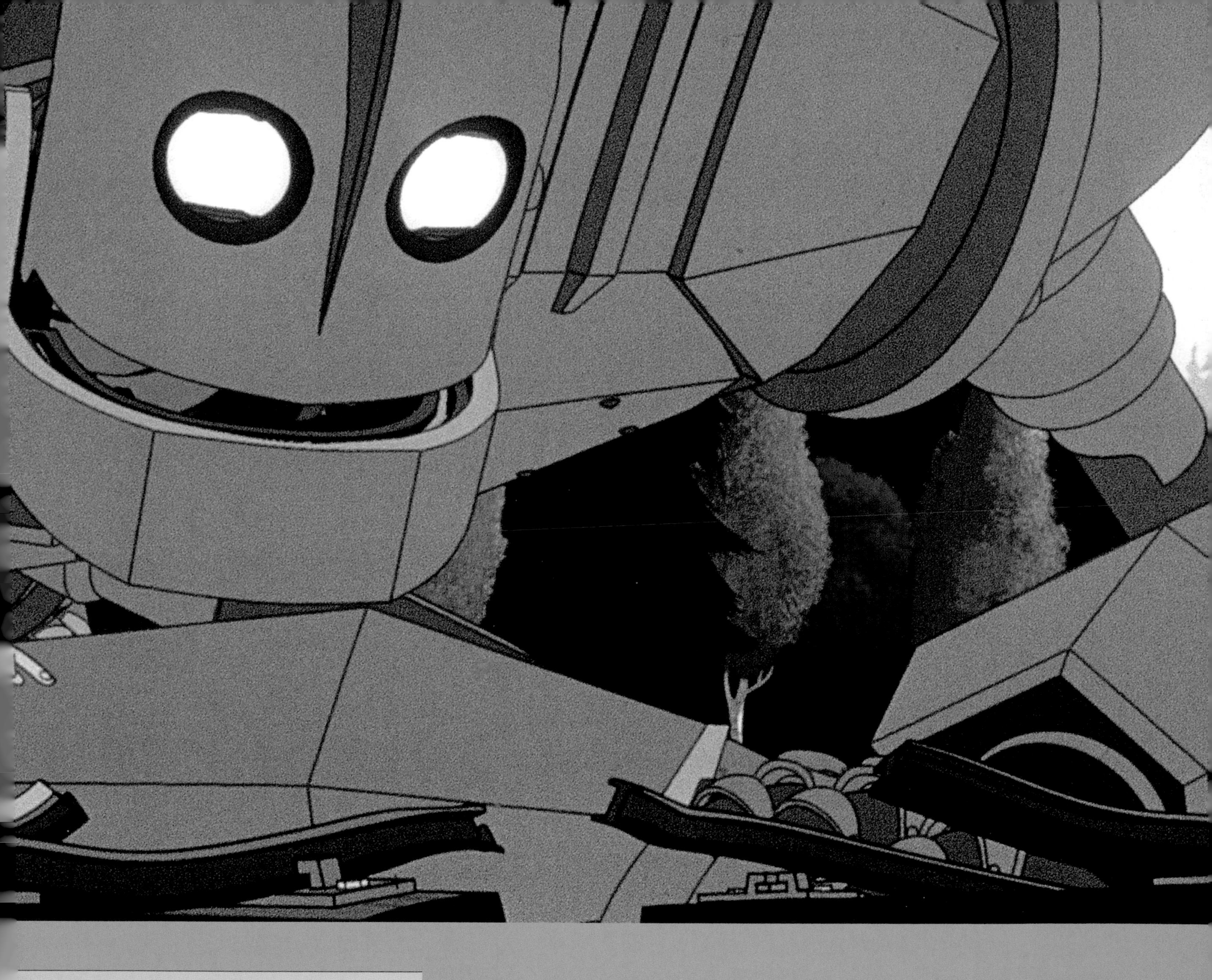

OPPOSITE: Marvin the Paranoid Android, from the original and best screen adaptation of Douglas Adams' *Hitchhiker's Guide to the Galaxy.*
ABOVE: one boy and his robot. The charming and beautifully realised animated movie *The Iron Giant,* based on the children's book *The Iron Man,* by Poet Laureate Ted Hughes.

4

Where it all
goes worng

When Robots Go Bad

Our need for/fear of technology illustrates one of the basic dichotomies of the 20th and 21st centuries.

As a race, we require increasingly sophisticated technology to fulfil our spiralling consumption of both basic subsistence items (such as shelter, light/heat and food) and business or leisure items. (Such is the rate of change in the last two decades some items, such as computers and mobile phones, have even—arguably—crossed over from business/leisure to subsistence.) But as much as we crave ever-more sophisticated technological cushions to pad out our bumpy existence, we also fear this *uber*-technology will ultimately strip away our livelihoods or, taken to the nth degree, our lives!

Consider the Luddites. Born out of the vast surge of industrialization that swept across Europe in the early part of the 19th century, the followers of the perhaps apocryphal Ned Ludd believed that the very machinery that was transforming the landscape of the British textile industry was simultaneously threatening their jobs and therefore their livelihoods, and so initiated an industrial revolution of their own, destroying the new mechanized looms wholesale. What those workers saw, um, looming on the horizon was not progress or streamlining or increased efficiency/productivity but their own redundancy, in both senses of the word. They feared that, unchecked, advances in science and technology would ultimately make them surplus to requirements. That same foreshadowing of future oblivion, of evolving technology overtaking our own evolution, is central to Karel Capek's play *R.U.R.* (Rossum's Universal Robots) and provides the nub of the fear element of many robot-based works of fiction or film since.

When, in the 1940s and 50s the idea of the robot, a mechanical being often in human form, designed to accomplish tasks previously reserved for human beings, was fully realized in the imaginations of philosophers, writers and filmmakers, it didn't take long for them to extrapolate and consider the eventual downside of such a progression. Not simply the somewhat tried and tested, 'what if they go wrong and turn on us?' scenario, but something altogether more insidious and disturbing. The question was posed: what happens to *us* when we are effectively replaced by robots? And, more chillingly, what happens if and when robots exceed their programming and decide that *we* are now surplus to requirements? What happens when robots go bad?

During the 'mad scientist' boom of the 1920s and 30s, bad (or at least wayward) robots were the norm. The *Frankenstein* model of a creature that lumbers off a drawing board somewhere and causes havoc was prevalent in much of early robot fiction and films, and now seems somewhat prosaic. Animator Nick Park paid due homage to these early clanking mechanical monstrosities in the *Wallace and Gromit* episode, *A Close Shave*. Preston, a robo-dog gone to the bad, exceeds his programming, and—having driven his creator to an early grave—turns sheep-rustler and dodgy dog food peddler, in the process framing hapless Gromit and nearly mulching cheese-loving Wallace and woolly Gwendolyn in an industrial meat grinder. As ever with Parks' work, it's the reverence for and respect of such bedrock storytelling standards that makes the pastiche so effective.

Asimov's robots of the 40s and 50s may have come with their built-in safeguards, but my own feeling is that his Laws of Robotics were more or less set up to be broken (or at least subverted). It was, I'm sure, almost a challenge for subsequent creative minds. The robots of Jack Williamson's novel *The Humanoids* represent the cusp of a sea change in the way authors and screenwriters saw robots could 'think'. Determined to obey their prime directive, the robots conclude that man is incapable of not slaughtering his fellow man wholesale and therefore decide to remove free

LEFT: robo-dog Preston from Nick Park's third *Wallace & Gromit* feature, *A Close Shave*. The script was co-written by Bob Baker, who had previously created Doctor Who's robotic assistant, K-9.

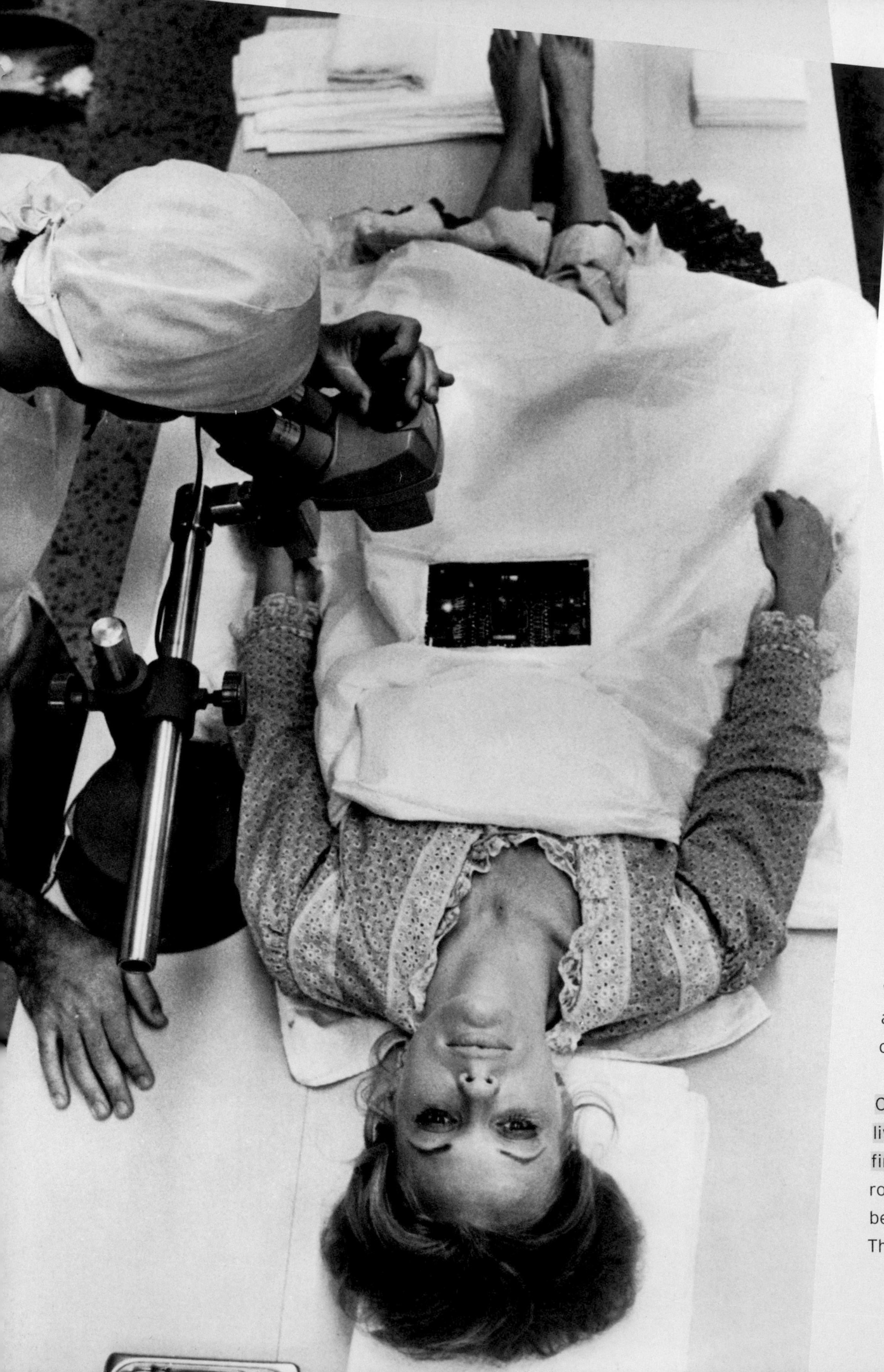

will entirely from the equation. Inevitably, this led to greater and greater flexibility in the overall structure and reach of what the robot story could be. Robots, in this context, are like a mirror for our own deepest fears and anxieties: what we create will ultimately destroy us.

Leaving aside for a moment the extraterrestrial threat, for me there are two main 'bad robot' scenarios. The first is epitomised by Michael Crichton's *Westworld* and the second by James Cameron's *The Terminator* (and its sequels). In *Westworld*, apparently inspired by a visit author and screenwriter Michael Crichton made to Disneyland and his first exposure to 'The Pirates of the Caribbean' ride, we have the ultimate theme park: a place where those with money can indulge their most escapist (adult) fantasies. Within the borders of the resort, Delos, are Western World, Roman World and Medieval World, recreations of each bygone era populated entirely by lifelike robots. Human guests interact with, fight with, even sleep with the robots, all of which activity is overseen by human coordinators and technicians in an atmosphere-controlled operations room deep under the resort.

Of course, there are safeguards. In Western World, live ammunition is the norm, but the guns won't fire on anything with a body temperature. The robots are programmed never to harm a human being, and always to lose in a fight (be it fist or gun). The fights themselves are carefully scheduled and stage-managed by the staff. As the synthesised Tannoy voice during the holidaymakers' induction drones repetitively, "nothing can go wrong." But, of course, it can... and does. The focus of the film, unsurprisingly, is Western World, and we follow the fortunes of holidaying lawyer Peter Martin (Richard Benjamin) and businessman John Blane (James Brolin) as they booze and brawl and carouse their way through a week in the pseudo frontier town. Martin is challenged to a gunfight by a nameless, black clad robotic gunslinger played by Yul Brynner and Martin duly shoots and kills the gunslinger, all according to the strict Delos 'script'. A second run-in ends with the same outcome. But while the gunslinger is off being repaired (in the process receiving a "routine upgrade"), an upsurge in small breakdowns and malfunctions starts to plague the resort(s). A robotic snake bites Blane, a hitherto unthinkable scenario.

Then, in one of the movie's most shocking sequences, Martin is challenged again by the gunslinger, but this time it's Blane who steps up to accept... and is promptly shot dead! The coordinators frantically try and shut things down—but it's already too late. Western World, Roman World and Medieval World all start to experience mass robotic malfunctions, resulting—inevitably—in the deaths of more and more guests. The robots are now running the show, grinding down their remaining stored power in an orgy of blood and carnage. No reasons are ever given for the robots' collective spiral into anarchy, but the subtext seems to be that the technology itself has become so advanced (in some cases, the robots are built by other robots) that it has somehow evolved to a next level of its own accord, and the human minds that began the process in the first place are now struggling to catch up. Or maybe Yul Brynner's gunslinger just got tired of getting shot by a lawyer!

Either way, *Westworld* isn't simply the old 'robot run amok' scenario dressed up new, though of course it has that basic heritage.

OPPOSITE: malfunctions start to multiply in the Delos resort and the technicians are at a loss to explain it: a scene from *Westworld*.
ABOVE: Yul Brynner's robotic gunslinger from *Westworld*. The black outfit was a visual reference to his role as Chris in *The Magnificent Seven*.

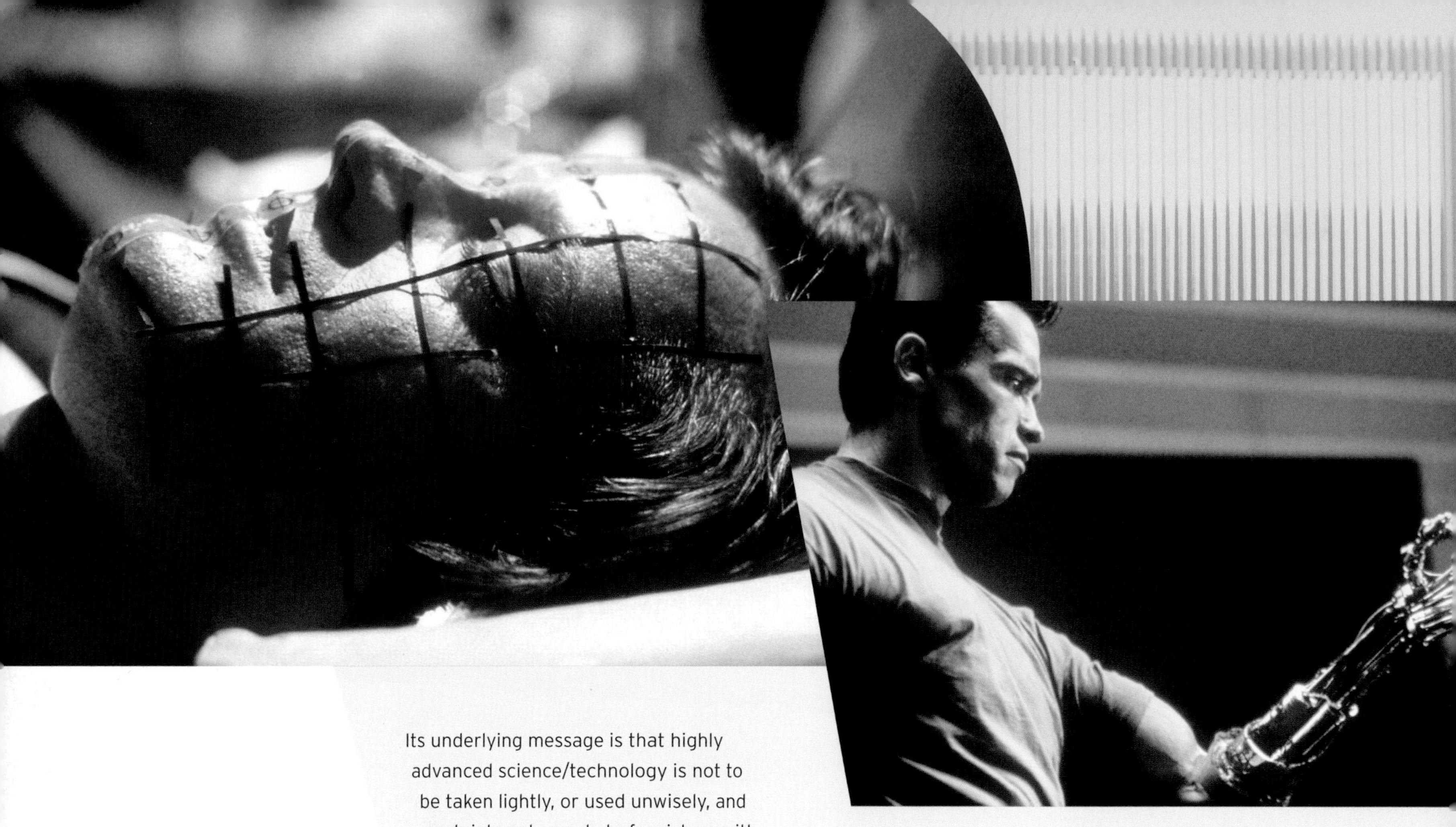

ABOVE, LEFT: Peter Fonda's features are digitally mapped in *Futureworld*. This tale of robotic doppelgangers was generally not well received by critics.
ABOVE, RIGHT: Arnold Schwarzenegger as the T-800 Terminator from James Cameron's techno-shocker *The Terminator*.
OPPOSITE: iconic film poster for *The Terminator*.

Its underlying message is that highly advanced science/technology is not to be taken lightly, or used unwisely, and certainly not merely to furnish us with idle amusements. It's a long, painful process, a struggle to understand an entirely new species, and it's one we rush at our own distinct peril. Crichton would later explore similar themes, only with genetically recreated dinosaurs, in *Jurassic Park*. In 1976, a less-well received sequel, *Futureworld*, was released, starring Peter Fonda and Blythe Danner, the story of which revolves around a plot by the owner of Delos to replace world leaders with clone robots. Yul Brynner's gunslinger is the only returnee from the first film, albeit in a bizarre 'erotic' dream sequence cameo. Though not without its moments, including some of the earliest digital compositing and CGI, *Futureworld* fails to recapture the stark, chilling simplicity of *Westworld*, miring itself in a post-Watergate conspiracy-style storyline (of which there were many at the time) that frankly creaks like an old, rusty bicycle wheel. Subsequently, a short-lived CBS TV series *Beyond Westworld*, starring James Wainwright, failed to find a sustainable audience.

Fast-forward to 1984, and a relatively low-budget sci-fi/action thriller from an almost unknown writer/director called *The Terminator*, was released without much fanfare or expectation. In terms of the 'bad robot' sub-genre, it was positively groundbreaking, and went on to become a hugely popular, multi-billion-dollar franchise, spawning two sequels, a 3-D show at the Universal Studios theme park, a TV series, novels and comic books, not to mention a new trilogy of movies starring Christian Bale. Whereas *Westworld* was all about our vanity and innate hedonism, the quest for 'bigger, better, faster,' *The Terminator* was underpinned by rampant paranoia, particularly as relates to defence and armed response, even more relevant in the post 9-11 security-conscious, Department of Homeland Security-heavy world we live in today.

The Terminator, played by Arnold Schwarzenegger, is a cybernetic organism, one of many created by a highly advanced and self-aware computer system known as Skynet, sent back in time to murder John Connor, the leader of the human resistance (to the machines' rule) in the future... before he was even conceived. The twist, among many, is that Skynet was originally a Department of Defence super-computer, its abiding purpose to protect America from its enemies. Unfortunately, Skynet had its finger firmly on the (nuclear) trigger, and as it gains sentience, it coldly and succinctly concludes that it would be better off without humankind (as a whole) and instigates an apocalyptic nuclear confrontation that comes to be known as Judgment Day.

The Terminators themselves are skeletal machines wrapped in living tissue, which acts both as a disguise (the Terminators are used for infiltration above and beyond their general hunter/killer remit) and facilitation (we learn that only living tissue can be sent back through time). Arnie's T-800 Terminator is everything a good bad robot should be: brutal, uncompromising, direct, unswerving,

ABOVE: Robert Patrick as the "mimetic poly-alloy" T-1000 from *Terminator 2: Judgment Day*. Patrick parodied his role in *Wayne's World*. **OPPOSITE, TOP:** massed ranks of endoskeletal Terminators from *Terminator 2: Judgment Day*. **OPPOSITE, BOTTOM:** the surprisingly badass Maximilian from Disney's *The Black Hole*.

focused entirely on its mission objectives, devoid of emotion or empathy, coldly logical—a guided missile with legs. There's no attempt by the filmmakers to put the T-800 through any kind of character arc or impose a moment of revelation. He/it simply is: a machine, as self-insightful as your average microwave oven. As Michael Beihn's character, Reese (a human protector sent back through time by John Connor to counter Skynet's retroactive strategy) puts it to the T-800's target, Sarah Connor: "It can't be reasoned with. It doesn't feel pity or remorse, or fear. And it absolutely will not stop, ever, until you are dead."

Though the real villain of *The Terminator* is undoubtedly Skynet, it's Arnie's implacable, un-emoting T-800 (a perfect bit of casting) that sears into the memory and ultimately carves out an enduring niche in this book's grand hall of robotic fame. It's made all the chilling by the fact that this nightmare nemesis is of our own creation, border paranoia personified. Once again, we are the

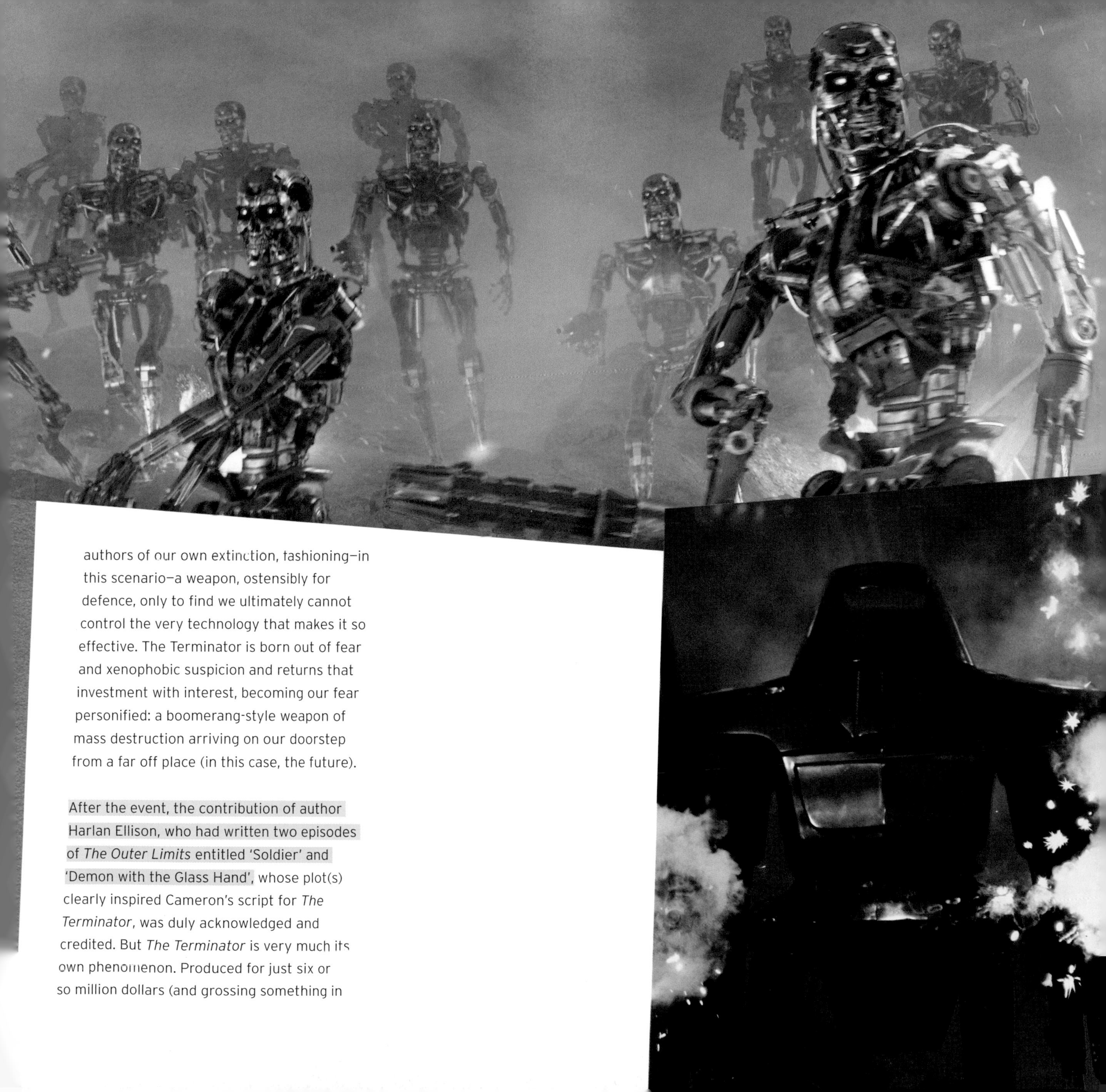

authors of our own extinction, fashioning—in this scenario—a weapon, ostensibly for defence, only to find we ultimately cannot control the very technology that makes it so effective. The Terminator is born out of fear and xenophobic suspicion and returns that investment with interest, becoming our fear personified: a boomerang-style weapon of mass destruction arriving on our doorstep from a far off place (in this case, the future).

After the event, the contribution of author Harlan Ellison, who had written two episodes of *The Outer Limits* entitled 'Soldier' and 'Demon with the Glass Hand', whose plot(s) clearly inspired Cameron's script for *The Terminator*, was duly acknowledged and credited. But *The Terminator* is very much its own phenomenon. Produced for just six or so million dollars (and grossing something in

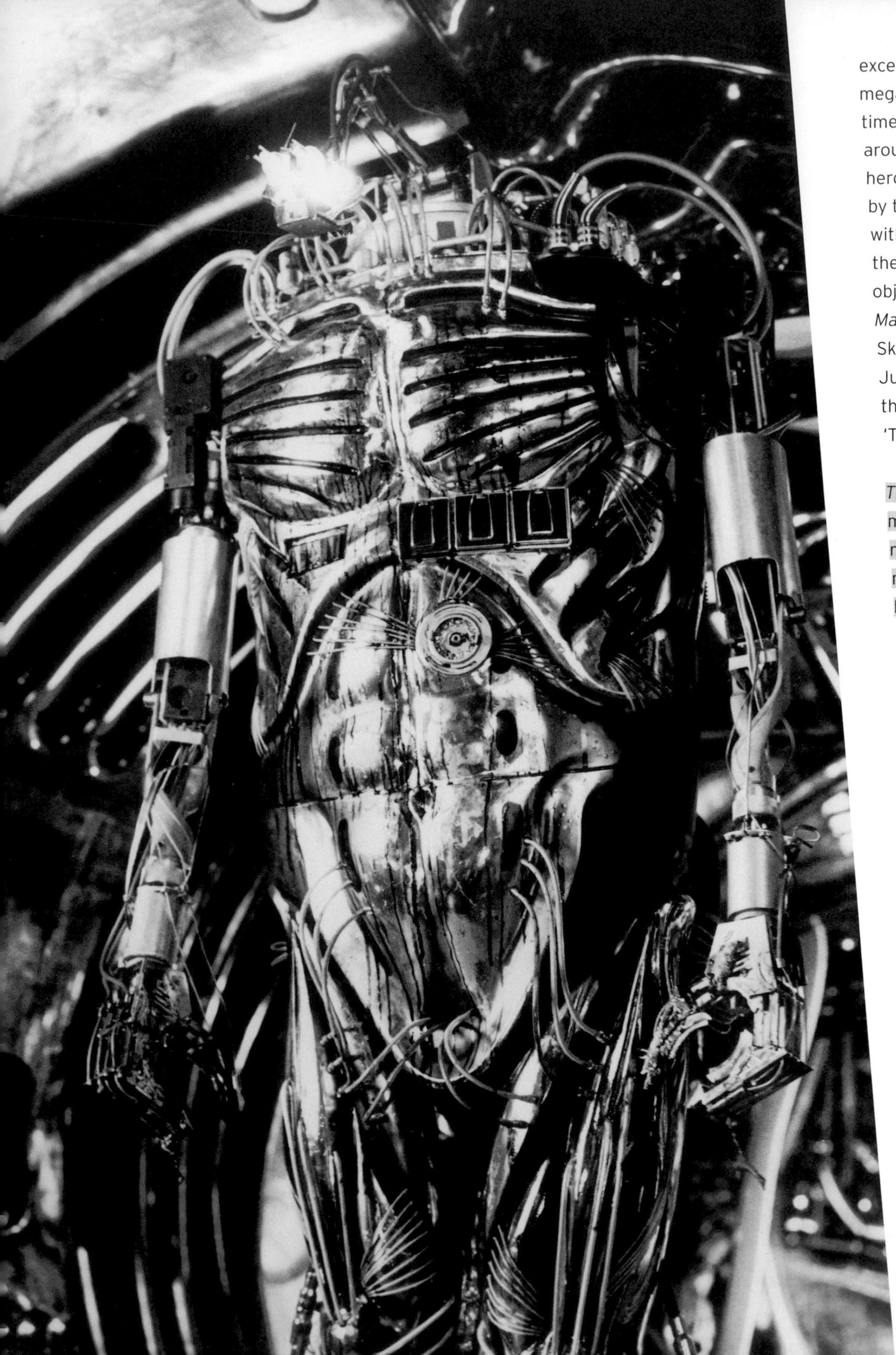

excess of $80 million worldwide to date), it made a megastar of Arnold Schwarzenegger and so, by the time the sequel—*Terminator 2: Judgment Day*—rolled around, it became necessary to cast Arnie in a more heroic light. This time around, Arnie (reprogrammed by the human resistance) becomes the protector, with Robert Patrick's liquid metal T-1000 providing the requisite unstoppable force to his immovable object. A second sequel, *Terminator 3: Rise of the Machines*, joins the dots between the evolution of Skynet/Cyberdyne Systems and the instigation of Judgment Day, the 'bad' Terminator role going, this time around, to Kristanna Loken (as the T-X or 'Terminatrix').

The Terminator franchise in no way has the monopoly on mad, bad and dangerous to know mechs. Considering he's in a Disney-produced movie, namely *The Black Hole*, Maximilian is one badass robot! At one stage, the hulking one-eyed 'servant' of Doctor Hans Reinhardt—a space explorer missing for 20 years and rediscovered on the periphery of a black hole's event horizon by the crew of exploratory ship, the *USS Palomino*—brutally eviscerates Anthony Perkins' character, Dr. Alex Durant, with whirling hand blades. And when, in the movie's climactic scenes, Reinhardt is trapped for all eternity within Maximilian's metal body, their new home on the far side of the black hole appears to be Hell, making Maximilian... the Devil? Well, maybe, there's a Faustian theme throughout the movie. Anyway, if so, ol' Max is a strong contender for *the* baddest robot of all time!

Hector, in the 1980 movie *Saturn 3*, is a case of nurture over nature. The prototype Demi-God (really, that designation doesn't bode well, does it?) series robot acquires the homicidal and downright lustful tendencies of deranged space pilot Benson (Harvey Keitel), via a direct feed programming

from his brain, thereby threatening the otherwise Eden-esque idyll of hydroponics scientists Major Adam (Kirk Douglas) and Alex (Farrah Fawcett). Hector is soon rampaging around the Saturn 3 space station, honing his murderous skills on the pet dog before graduating to the equivalent of patricide by killing his father/creator Benson. Thereafter, a lot of sub-*Alien* running around in dimly lit corridors ensues, until finally Adam sacrifices his life to stop Hector once and for all. Ultimately, *Saturn 3* falls somewhere between two stools, neither lurid enough to be an exploitation movie or po-faced enough to be taken seriously. But Hector does look good, all seven feet or so of external innards and tiny vestigial head on extended neck. Six out of ten for Hector, nil points for *Saturn 3*.

Corporate bad robots almost deserve a sub-sub-categorization. The first and perhaps best of these was Ash, from Ridley Scott's *Alien*. For the longest time, we don't even know Ash is an artificial being. He appears to be just another crewmember aboard the ill-fated space freighter *Nostromo*. But Ash has a whole other agenda, namely bagging a killer alien for the company, and if he has to sacrifice every other member of the crew to get it back to Earth so be it. Exhibiting even less emotion than the ship's computer, 'Mother', Ash sets about killing Ripley (Sigourney Weaver) with a rolled up wadge of paper, before having his head knocked off with a fire extinguisher. But you can't keep a bad corporate robot down, and spluttering

OPPOSITE: horny Hector from *Saturn 3*, one of the most visually striking cinematic robots of all time. Shame about the film! **ABOVE:** "Dick, I'm very disappointed." OCP's president expresses his frustration to Dick Jones (Ronny Cox) when their ED-209 droid brutally shoots and kills an employee. From *Robocop*.

something that looks like curdled milk, Ash (now just a head) tells the crew they're all going to die. What a trooper! Then there's Enforcement Droid Series 209 (ED-209 for short), a squat, machine-gun armed pitbull of a droid designed and built by O.C.P. (Omni Consumer Products) for urban pacification, and subject to routine and bloody malfunctions, from Paul Verhoeven's *Robocop*. Low on IQ, but big on zero tolerance. By the way, Robocop himself (as played by Peter Weller) doesn't make the cut for this book as, his name not withstanding, he's not strictly a robot, falling more into the cyborg category (ie. part human, part machine).

Not exactly a corporate robot, but verging on one, is SID 6.7, as played by Russell Crowe in *Virtuosity*. S.I.D. (Sadistic, Intelligent, Dangerous) starts life as a virtual reality construct, part of a training program for police officers developed by the Law Enforcement Technology Advancement Centre. Programmed with over 150 separate criminal personalities, from mass murderer to terrorist and pretty much everything in-between, SID tricks his way out of cyberspace and into a nano-machine android body, embarking on an all-too tangible killing spree. Reinstated cop Parker Barnes (Denzel Washington), who's already faced SID in the virtual environment and (just about) survived, is given the task of stopping SID for real.

Bad robots proliferate in Michael Crichton's *Runaway*, as Sgt. Jack Ramsey (Tom Selleck) investigates bizarre murders perpetrated by otherwise docile domestic robots. But it's the acid-injecting robo-spiders created by evil genius Dr. Charles Luthor (Kiss frontman Gene Simmons) that leave the most lasting, arachnophobic impact as they skitter here and there, up walls and across ceilings, and—in one notable scene—converge en masse on Ramsey high up on an elevator platform.

OPPOSITE: Tom Selleck in *Runaway*, menaced by Gene Simmons' robotic spiders.
RIGHT: film poster for *Chopping Mall*, a low budget but decently inventive slasher movie originally released under the title of *Killbots*.

The low-budget 1990 sci-fi thriller *Hardware*, directed by Richard Stanley, has a certain post-apocalyptic raw energy as killer combat droid M.A.R.K. 13, a reference to the Gospel of Mark, which contains the phrase "no flesh shall be spared", first reassembles itself from spare parts and then terrorizes Stacey Travis and Dylan McDermott. But the film, lean and mean as it is, is marred irrevocably by the fact that the script was cynically plagiarized from a short story in cult British sci-fi comic *2000AD* called 'SHOK', by Steve MacManus and Kevin O'Neill, and only subsequently—following legal action—credited as such on prints of the film.

Also of note, however brief, are *Evolver* and *Chopping Mall*. In the former, a 'Sci-Fi Channel original movie', teenager Kyle Baxter (Ethan Embry) defeats the virtual reality Evolver robot in a video game competition and his prize is a real, nuts and bolts version of the same. True to its name, the robot starts to evolve, and the 'game' soon becomes deadly reality. In the latter, teenagers are hunted down and killed by robo-security guards called 'Protectors' in the state-of-the-art Park Plaza shopping mall. The film was originally released as *Killbots* but performed poorly at the box office. The subsequent re-titled/re-released version performed better, especially on its DVD release. In what is a reasonably routine teens-in-threat movie, there are a couple of nice touches: the tag line, "where

TOP, RIGHT: Godzilla versus the mighty Mechagodzilla from *Godzilla vs the Cosmic Monster*.
BOTTOM, RIGHT: Steed and Mrs Peel meet the robotic Cybernauts in *The Avengers*.

shopping costs you an arm and a leg!" (though actually there's no dismembering as such), and the fact that—after they've killed you—the Protectors say, "have a nice day."

But bad robots aren't limited to the big screen, oh no, TV's had its fair share of them too, including the (alien-created) Cylons from the original *Battlestar Galactica*, the robot Bigfoot from *The Six Million Dollar Man*, Data's evil twin Lore from *Star Trek: The Next Generation*, the Repli-Carter from *Stargate SG-1*, the 'Christmas Angels' from the 2007 *Doctor Who* Christmas Special (and, for that matter, the robots from the classic Tom Baker Doctor Who story, 'The Robots of Death' and the clockwork androids from 'The Girl in the Fireplace', and... well, I could go on), the Cybernauts, recurring robotic villains from the Steed and Mrs Peel era of *The Avengers* and the later *New Avengers* and the evil version of the Robot B-9 in *Lost in Space*, from the season three episode 'The Anti-Matter Man'. In fact, bad robots proliferate in *Lost in Space*, the tiny robots in 'The Mechanical Men' and I.D.A.K. (Instant Destroyer And Killer... gotta love whoever named that robot!) from 'Revolt of the Androids'. Bender, from *Futurama*, also broadly fulfils the criteria for a bad robot, but rest assured he gets some love elsewhere in the book.

Our final sub-sub category is extraterrestrial bad robots, and they don't come much bigger and badder than Megatron and his Decepticons, from the *Transformers* animated TV series and 2007 Michael Bay movie. Megatron—and a whole bunch of other Transformers—also get their column space elsewhere in the book. The biggest surprise I got when researching this book was the marked paucity of bad robots in the *Star Wars* universe. Strange, when you consider the sheer number of droids and so forth running around. There's really only, in any prominent shape and form, General Grievous and IG-88, a soft-spoken but ruthless bounty-hunter. C'mon, Lucas, pull your finger out! And last but not least, possibly my favourite bad mech of all – Mecha-Godzilla, from (originally) *Godzilla vs. Mechagodzilla*. So utterly bad (in every sense of the word) he's good, Mecha-Godzilla mixed it up with Godzilla and others in any number of Japanese creature features. Of course, what exactly Mecha-Godzilla says about the human condition is anyone's guess!

ABOVE: the far-from-festive Christmas Angels in the 2007 *Doctor Who* Christmas Special, 'Voyage of the Damned'.

5 Robots With Dangerous Curves

Sexy yet deadly:
perfection with
a price

It's perhaps the strangest, and yet-conversely-most easily graspable of all the many pop culture variations on a robotic theme.

Perfect, sexy (usually female) automatons: the ultimate in male fantasy/wish fulfilment. Women you literally mould—according to your own unique specifications, with a slew of extra features that would put a BMW X5 to shame—and, more importantly, you control. Or not. As a whole host of sexy yet deadly robots have demonstrably shown over the last eight or nine decades of fiction and film, chromium curves come with a price. Beneath the all-too-perfect veneer of "fembot" submission lurks colossal power and simmering aggression, which is almost always directed back against the male of the species. Sex and danger... an age old, potent and heady cocktail! Let's jump right in...

The term fembot immediately brings to mind the nipple-gunned robotic agents of Doctor Evil from *Austin Powers, International Man of Mystery* (and its sequels), but sexy or siren robots have been a staple of cinematic tradition for a lot longer than the last decade or so. Indeed, one of the very first fully-fledged screen automatons of all was the 'fake' Maria, from Fritz Lang's *Metropolis*, a rabble-rousing robotic seductress putting it all out there in an effort to foment revolution among the downtrodden masses of Lang's future world, bumping and grinding out her inflammatory speech with all the calculated tease of a stripper. However, quite why robot Maria is so committed to this incendiary course of action isn't clear when you see the savagely trimmed English-language version.

In the original German cut of *Metropolis*, a whole subplot (since lost) involves a woman, Hel, who was loved by the robot's creator, Rotwang, but who married Joh Fredersen, the current ruler of the city, instead. Hel dies giving birth to Fredersen's son, but remains forever the object of Rotwang's obsession, her statue part of a permanent monument erected at his house. Rotwang reveals triumphantly to Fredersen that the robot will be given the appearance of Hel, and so she will 'live' again. (A plot that itself harks back to the old Greek legend of Pygmalion, who fell hopelessly in love with a statue that was then given life by the Goddess Aphrodite.) Fredersen urges Rotwang to give his robot the appearance of Maria, rather than Hel. At first, Rotwang is resistant, but ultimately realizes he can use robot Maria to drive a wedge between Fredersen and his son, Freder, who is madly in love with (the real) Maria.

This whole sequence of events, cut for the U.S. release, goes a long way to making sense of the various motivations in the film, and explains why robot Maria is quite so voraciously determined to foster revolution and chaos in the under-city. In fact, in another cut/lost scene, Rotwang specifically programs the Maria-to-be with orders to "destroy Joh Fredersen... and his city," while (simultaneously) Freder listens to a sermon/prophecy that warns of a coming apocalypse that will announce itself in the form of a sinful woman. Once you know all this, it's easy to see just how neatly 'fake' Maria fits the eventual standard-operational spec of the fembot, at once beguiling and destructive, a wolf in designer sheep's clothing.

After *Metropolis*, which—thanks in no small part to the now somewhat nonsensical story—had a less than rapturous reception from audiences and failed to deliver at the box office, the whole fembot concept pretty much went into a kind of suspended animation, until, in 1975, when *The Stepford Wives* (based the novel by Ira Levin) rounded off the blocky edges and gave robots back their hourglass figures. Though, I should note, that over on the small screen Julie Newmar (she of Catwoman fame) had already been raising male temperatures as Project AF-709 in *My Living Doll*, a CBS sitcom that ran from September 1964 to March 1965. Newmar's Amazonian robot, who can be turned on and off by pressing a beauty spot on her back (!), is a U.S. Air Force science project, entrusted into the hands of psychiatrist Dr. Bob McDonald (or MacDonald, the stencilled sign on his office door changes at one point)

ABOVE: fembot Vanessa Kensington (Elizabeth Hurley) reveals more than super-spy Austin Powers bargained for in *Austin Powers: The Spy Who Shagged Me*.
RIGHT: how to turn a woman on. Inventor Dr Carl Miller (Henry Beckman) presses the right button/beauty spot on AF-709 (Julie Newmar) in the CBS sitcom *My Living Doll*.

ABOVE: perfect partners. The robotic spouses from *The Stepford Wives*, based on the darkly comic novel by Ira Levin.
RIGHT: Sean Young as Rachel, the epitome of artificial sophistication, as created by the Tyrell Corporation in *Blade Runner.*

by her creator, Dr. Carl Miller when he's reassigned to Pakistan. McDonald (played by Bob Cummings) conducts something of a social experiment, intent on inducting Rhoda (as he names AF-709) into society, having concluded that she is essentially the perfect woman. As McDonald puts it in the pilot episode, 'Boy Meets Girl', she "does as she's told... reacts as you want her to react... and keeps her mouth shut." Hm. Mired as it is in 60s sexist attitudes, it's hard to love *My Living Doll*, but Newmar plays it admirably straight, and its legacy, such as it is, is the phrase, much used since in the robot genre, "does not compute."

Back to *The Stepford Wives,* in which the aforementioned male wish-fulfilment is taken to a whole new level. Written by Ira Levin and adapted for the screen (first time around) by William Goldman, the story takes us to Stepford, Connecticut, a model, idealized American town complete with model, idealized American wives, all perfectly turned out, subservient and with little on their minds except their domestic/household duties. New Yorker Joanna Eberhart, played by Katherine Ross, her husband and two children, are the new family in town, and Joanna quickly surmises all is not well in Stepford, despite the idyllic sheen. She makes friends with another newcomer, but soon she too has been transformed into a squeaky clean housewife. Now more than simply suspicious, Eberhart investigates the cliquey Men's Association, of which her own husband has become a member, only to come face-to-face with her own robot double and the man behind the plot to replace all the town's wives with perfect, obedient artificial women. She battles her own duplicate but, chillingly, it is the robot Eberhart who—in the final scene—we see dutifully shopping at the local supermarket.

No happy endings here, just a bleak and damning portrait of a male-dominated culture, where women are merely beautiful objects that exist primarily to serve. Of course, at the time, society was changing, attitudes becoming more liberal, and *The Stepford Wives* was a thinly veiled attack on those kind of outdated sexist attitudes, but you wonder if, thirty or so years later, those attitudes have really changed. Perhaps, given the proliferation of idealized sexy robots since, they've just gone further underground. Such pontifications aside, *The Stepford Wives* really crystallized the notion of a fembot as its own distinct sub-genre, and the generic term 'Stepford Wives' has very much entered the more popular lexicon, as a byword for air-headed, too perfect, consumer-product obsessed women.

The book/film spawned a number of inferior sequels, including *Revenge of the Stepford Wives* (also based on a novel by Levin), *The Stepford Children* and, mixing things around somewhat in a vague attempt to present a more balanced picture, *The Stepford Husbands*. In 2004 the film was remade with Nicole Kidman and Bette Midler, and turned into an uneven comedy-drama. Plagued by plot holes, largely due to a last minute, radically re-thought ending in which the 'robots' turn out to be the real wives with implanted control chips, despite having been seen to dispense cash from their mouths, crackle with electricity and so forth, *The Stepford Wives* remake is perhaps best sidestepped.

Better by far to jump straight to Ridley Scott's *Blade Runner*, which is positively awash with fanciable fembots. Of the four rogue replicants (five if you go for the whole 'Harrison Ford's character, Deckard, is one too' theory), artificial beings designed for off-world hard graft, two are female: Pris, played by Darryl Hannah, and Zhora, played by Joanna Cassidy, and icily beautiful Sean Young plays the 'more human than human' Rachael, the Tyrell Corporation's most sophisticated model. Now these are fembots with both poise and punch, from Rachael's demure, doll-like indifference to Pris' gymnastic, thigh-clenching animosity,

determined to shake off the racial shackles imposed on them by their creators. Here, it's not so much about sexual identity as identity itself. The theme underpins much of *Blade Runner*. Who is who? Who is what? And, ultimately, should it/does it matter at all? We are who we are and all the rest is just someone else's branding.

Of course, that doesn't stop embattled, burned out Blade Runner (future cop/bounty-hunter) Harrison Ford 'retiring' both Zhora (shooting her in the back as she plunges headlong through a series of plate glass shop windows) and Pris (blowing a big hole in her, even after she's stuck his head between her thighs) and running off into the wild blue yonder with Rachael. It's a happy ending of sorts, at least for one fembot, even if it was tacked on to the original release and later subtracted for the Director's Cut and so on. Demure wins out over dangerous in *Blade Runner*, which perhaps is some sort of life lesson in itself. We start out, in our youth, seeking out-there sexual thrills but—older, wiser—settle down with a safer, more serene life partner. Or it's just a bunch of hot 'bots making the bleak future a little brighter. You decide.

If *Blade Runner* is the sublime of the fembot sub-genre, *Weird Science* is the ridiculous. Though, that's not to say *Weird Science* doesn't have its own, albeit shallow, merits.

It has. It's just a very, very different beast. Now, I should stress that I know the perfect female dreamed up by teenage nerds Gary Wallace (Anthony Michael Hall) and Wyatt Donnelly (Ilan Mitchell-Smith) and made flesh by Kelly LeBrock isn't, strictly speaking, a robot, but she is designed wholly in-computer and given life by a *Frankenstein*-inspired bolt of lightning, and that's good enough for me. Directed by John Hughes, of *The Breakfast Club* and *Home Alone* renown, *Weird Science* is mostly an excuse to showcase the curvaceous Kelly LeBrock in next to nothing and titillate teenage boys, but by referencing *Frankenstein* (or at least its filmic sequel, *The Bride of Frankenstein*) and generally and cheerfully lampooning the whole fembot phenomenon it does merit a nod from me. Other than that, *Weird Science* is pretty much boys make girl, girl makes men of boys (though not actually by bedding them). But if you fancy switching off your brain and being amused (in a sledgehammer type way), I can think of worse candidates.

Moving swiftly on, the late 80s and early 90s yielded only a clutch of titanium temptresses, including Cherry 2000, from *Cherry 2000*, Eve-8, from *Eve of Destruction*, Alsatia Zevo from *Toys* and Casella 'Cash' Reese, from *Cyborg 2*, none of them exactly prime examples of the phenomenon but each more than filling the various fembot criteria necessary for at least a mention.

Cherry 2000 is interesting in that the titular (*note to self: must stop using that word, especially in this chapter!*) Cherry is both intrinsic to the plot and, at the same time, only a very peripheral character. It's the future, and romance is well and truly dead. Dates are arranged, in-toto, up front, right down to the nitty-gritty and complete with forms and contracts. Our protagonist, Sam Treadwell (David Andrews) opts, instead, for a Cherry-2000 pleasure robot (played by Pamela Gidley), only—as we join the film—he's rather worn her out. A particularly energetic sexual encounter in one of those movie-only oceans of soapsuds leads to Cherry blowing a fuse. Unable to find a replacement body, the Cherry-2000 line long curtailed, Treadwell resorts to a journey into the vaguely post-apocalyptic badlands known as Zone-7, accompanied by flame-haired

OPPOSITE: poster for *Weird Science*, featuring CG construction made flesh Kelly LeBrock.
ABOVE, LEFT: poster for *Eve of Destruction*, re-titled as *Priorità Assoluta* (Absolute Priority) for its Italian release.
ABOVE, RIGHT: Pamela Gidley as the (somewhat knackered) sex toy Cherry 2000 in *Cherry 2000*.

tracker E. Johnson (Melanie Griffiths), where, he's told, he'll find a factory full of Cherrys. In the process, amidst a lot of sub-*Mad Max* action, he comes to realise, maybe from having Melanie Griffiths sat in his lap while he drives a car, that flesh and blood babes are best. Aww. There's a fun scene where Treadwell is offered a range of alternate bodies in which to insert... wait for it, Cherry's memory chip, and among them are Gort (from *The Day the Earth Stood Still)* and Robby (from *Forbidden Planet*). Oh, by the way, the music, by Basil Poledouris, is great. Best thing about *Cherry 2000* by a country mile.

Eve of Destruction is a by-the-numbers, rogue mech *Terminator* rip off that pre-empts that franchise's own 'Terminatrix,' the T-X, by a decade or so. Eve-8 is a female (military) robot built in the image of her creator, Dr. Eve Simmons (Rene Soutendijk, who plays both roles) and then carelessly let loose among the general population. A bungled bank robbery activates her in-built self-defence mechanisms and destruction ensues. It's down to super S.W.A.T. operative Jim McQuade (Gregory Hines) to stop Eve-8 before she detonates, taking half the city with her. That's right, Eve-8, for no particularly good reason other than as a plot contrivance, has a nuclear (self-destruct) device built in. And that's about it. Well, there's more, but—honestly—I just hit the fast forward button.

In *Toys,* a strange, uneven oddment of a film from director Barry Levinson, Joan Cusack and Robin Williams play sister and brother Alsatia and Leslie Zevo. In the climactic cuddly toy-versus-combat toy battle, Alsatia is revealed to be an android, manufactured by Leslie's father, a master toymaker, to comfort Leslie following the early death of his mother. *Cyborg 2* can, I feel, be encapsulated with equal economy of words. Angelina Jolie is Casella 'Cash' Reese, a cyborg designed to detonate (she carries something called 'Glass Shadow' in her veins/wires) when she orgasms (seriously). One of her (there are many) falls in love with her martial arts teacher, Colton 'Colt' Ricks and—helped by Mercy, another cyborg played by Jack Palance—they rebel against/escape from their corporate owners, Cash cheating detonation and living happily ever after with Colt, or at least until he dies of old age. *Cyborg 3*, by the way, is subtitled 'The Recycler'. You've got to love that kind of 'does what it says on the tin' honesty.

A proud tradition of androids in the *Alien* movie franchise was continued in *Alien: Resurrection.* Previous 'artificial humans' Ash (Ian Holm in *Alien*) and Bishop (Lance Henrikksen in *Aliens*) were joined in the third sequel by fembot Annalee Call (Winona Ryder). Having already had a bad android (Ash), a good android (Bishop), Call is somewhere in the middle. At first, she tries to kill Ripley (Sigourney Weaver), but only because she believes Ripley, her human D.N.A. now spliced with Alien D.N.A., will be used to create more Aliens. When, ultimately, she realizes Ripley is now committed to killing all the Aliens aboard the mercenary ship, *The Betty*, Call joins forces with Ripley.

Other notable filmic fembots include the aforementioned agents of Doctor Evil in *Austin Powers, International Man of Mystery.* Inspired by the fembots from the TV series, *The Bionic Woman*, these blonde, bouffant-haired robotic assassins would become a staple of the Austin Powers movies. In *Austin Powers, the Spy Who Shagged Me*, it's revealed that Vanessa Kensington (Elizabeth Hurley), the female lead from the first movie that Austin subsequently marries, is a fembot too. When, after Vanessa has just tried to kill him, Austin informs his boss, Basil Exposition (Michael York) of this startling fact, he learns that the ever-unruffled Basil knew all along. Austin, after the briefest moment of consternation, just shrugs this off. *Austin Powers in Goldmember* features Britney Spears as a fembot. Well, we all knew that!

At least *A.I. Artificial Intelligence* provided some artificial eye candy for the ladies in the shape of Jude Law as Gigolo Joe. Based on the Brian Aldiss short story, *Super-Toys Last All Summer Long* and originally developed for the screen by the late Stanley Kubrick, *A.I.* (co-written and directed by Steven Spielberg) focuses, primarily, on

OPPOSITE, TOP: Angelina Jolie kicks ass in *Cyborg 2.* Everyone has to start somewhere! **OPPOSITE, BOTTOM:** the proud tradition of artificial humans in the *Alien* movies is continued in *Alien: Resurrection* by Winona Ryder as Annalee Call. Better looking by far than either Ian Holm (*Alien*) or Lance Henriksen (*Aliens*).

David, an artificial boy termed a 'mecha'. David (Haley Joel Osment) goes to live with Henry and Monica Swinton, whose real son, Martin, has been afflicted by a rare disease and placed in suspended animation until a cure can be found. At first, David is loved, and—programmed with the ability to love—he loves his 'parents' back. But when Martin is cured and returned to the family home, events spiral out of control, driven by Martin's jealousy and David's self-protection program. The Swintons abandon David and he is forced to go on the run with Gigolo Joe, a male prostitute mecha framed for murder.

Gigolo Joe is all smooth shimmies, perfectly sculpted hair (the colour of which he can change with a flick) and a note-perfect line in seduction, his flattering patter and in-built, one-click romantic music reducing lonely ladies to mush (for financial remuneration). "I know women," Joe tells David when the two are thrust together after their escape from a 'flesh fair', and he certainly does. Jude Law plays Joe perfectly, everything about him sharp, sexy and loaded with the requisite amount of seedy illicitness. We meet—briefly—his female counterpart Gigolo Jane (Ashley Scott), but she's there and gone with the briefest "whaddaya know, Joe?" And I'm sure many men would love to know what Joe knows!

Of course, there are many other male screen robots that classify, according to taste, inclination, etc, as 'sexy'. Arnold Schwarzenegger in *Terminator,* Rutger Hauer in *Blade Runner*, Russell Crowe in *Virtuosity,* but largely those films (and their stars) fall more into other categories. However, perhaps Klaus Kinski from the film *Android* deserves a mention here, not least because he, and the film, weren't a neat fit elsewhere! This low-budget but intriguing sci-fi film from director Aaron Lipstadt involves eccentric scientist Doctor Daniel (Klaus Kinski), who is illegally creating humanlike robots

on a remote space station, among them his prototype/assistant Max-404 (Don Keith Opper) and his 'perfect woman' Cassandra One (Kendra Kirchner). When three escaped convicts, among them Maggie (Brie Howard), take refuge at the station Max-404 falls for Maggie and determines to escape the confines of the space station. The final twist comes when we learn Doctor Daniel is an android too! What's most interesting about *Android* is the film is presented totally from the androids' emotional point of view, making Max-404 and Co forerunners of *Blade Runner's* questing replicants.

More recently still, actress Kristanna Loken appeared as the T-X in *Terminator 3: Rise of the Machines.* Loken's 'Terminatrix', in a somewhat futile effort on the behalf of the writers and production staff to outdo Robert Patrick's T-1000 in terms of advanced abilities, can remote control other machines and morph her hand/arm into a variety of deadly weapons. But, after the sheer adaptability of the T-1000, the T-X comes across as more of a step back than a progression. Loken, though, is suitably remorseless and icy, managing to match Arnie (again the protector) in terms of sheer screen presence. The twist of having a

OPPOSITE: Jude Law as Gigolo Joe and Ashley Scott as Gigolo Jane strike a pose in Steven Spielberg's *A.I.: Artificial Intelligence.*
ABOVE: all rise! Kristanna Loken as the T-X, or Terminatrix, in *Terminator 3: Rise of the Machines.*

LEFT: Doyle (Brandy Ledford), Beka (Lisa Ryder) and Rommie (Lexa Doig) from *Andromeda*. Spot the human!

OPPOSITE: Tricia Helfer as Cylon number six from the triumphal recreation of *Battlestar Galactica*. Much more than just a pretty face!

RIGHT: *Mann & Machine*, the 80s cop/robot series starring David Andrews and Yancy Butler (she's the robot, by the by).

female Terminator was previously explored in the novels *T2: Infiltrator* and the comic book *Terminator One-Shot* (collected in *Terminator Rewired*), and re-visited in the TV show *Terminator: The Sarah Connor Chronicles*, in which the protector Terminator, Cameron, is played by Summer Glau.

Cameron also forms part of a fairly small group of TV fembots, which includes (above and beyond Julie Newmar's AF-709 from *My Living Doll*); Andromeda from *A For Andromeda*, a 1961 BBC production (and its 1962 sequel, *The Andromeda Breakthrough*); the Buffy-bot from *Buffy The Vampire Slayer*, a robotic likeness created on behalf of besotted vampire Spike and later standing in for the real Slayer (Sarah Michelle Gellar) following her 'death'; Valerie-23, from an episode of the newer, raunchier version of *The Outer Limits*; Rommie, from *Andromeda* (the robotic avatar of an entire star-ship), played by Lexa Doig; Doyle, a female android played by Brandy Ledford (from the same series) and Sgt. Eve Edison (Yancy Butler) from *Mann & Machine*, a short-lived NBC sci-fi/police drama. As a kind of 'Six Degrees of Kevin Bacon' interconnections sidebar, *Mann & Machine* also starred David Andrews as the human cop partnered with Yancy Butler. Andrews would later go to great lengths to recover his robotic, ah, soulmate in *Cherry 2000*.

However, any such small screen shortcomings are more than made up for by the revamped/ re-imagined version of *Battlestar Galactica* (2004-2009). Instead of being creations of an alien race, the robotic Cylons are of human origin, and when they ultimately rebel, a terrifying war ensues. Thereafter, a cagey truce is reached, and the Cylons retreat to a world of their own, appearing only for brief, perfunctory armistice summits. Forty years pass, and we learn that the Cylons have evolved, taken on human form. There are now twelve perfect Cylon-human likenesses, duplicated ad infinitum, among which are three females: number six, played by Tricia Helfer, number eight (who is also Lieutenant Junior Grade Sharon 'Boomer' Valerii aboard the Battlestar *Galactica*), played by Grace Park, and number three, played by Lucy Lawless. Not only are this trio the strongest and most proactive of the humanoid Cylons, they're also the most chimerical, juggling emotional issues such as love, motherhood, faith and compassion, evolving throughout the course of the four series (plus opening mini-series and one-off, *Razor*) into deeply complex, multi-layered individuals (despite the fact that each downloads into successive bodies).

Number six, all seductive slink and nonchalant capacity for colossal destruction is perhaps the most iconic of the trio, combining sex and death to truly knockout effect. Her strange relationship with human scientist Dr. Gaius Baltar, a narcissistic womanizer, forms the backbone of the series. Initially, he's the weak link in the human

defences, exploited by number six to get at the planetary codes that protect the human colony worlds from attack. But Baltar and number six's attachment remains curiously constant throughout, forcing both to re-examine their lives, beliefs and priorities, not to mention their overall place in the scheme of things. For the longest time, number six exists as a kind of internal voice in Baltar's head, cajoling, coercing and comforting in equal measure. It's a credit to Tricia Helfer, who plays number six, and the show's writers that the character is quite so multi-dimensional, instead of the standard sexy but lethal siren. 'Boomer' too is made to leap through incredible character hoops as she goes from deep cover Cylon agent to human sympathiser to mother of the first human/Cylon hybrid child.

So different from its 80s counterpart it's hard to consider them in any way interrelated other than in name, 'new' *Battlestar Galactica* is a sweeping, multi-faceted science fiction opus, one that breathes new life into all manner of dusty old standards, not least that of the fembot, consistently challenging all the genre preconceptions by tackling, often unflinchingly, real world issues and offering no easy, black and white solutions. It's interesting to note that in this new order of life, somewhere between human and robotic, it's primarily the women who are strong and focused, driving the quest for self-realization and global and spiritual growth, often rising above their own innate agendas. Finally, after much to-ing and fro-ing, the fembot had come of age.

Outside of films and TV, the fembot's influence is likewise pervasive. In 1979, Japanese artist Hajime Sorayama practically coined the phrase 'sexy robot' with his seductive and provocative paintings of metallic hued naked (or near to it) women. Then there's Heineken's sexy, dancing robot with its own inbuilt keg of beer, part of a recent advertising campaign, and the Philips/Nivea For Men 'robot skin' advertising films, plus Icelandic pop star Bjork's video for *All is Full of Love*, wherein two robotic Bjorks fall in love. The fembot walks a thin line between allure and absolute power, simultaneously tantalizing and terrifying. And we... just can't get enough!

Oh, and just a quick, last gasp mention for Rosie, from the Jetsons cartoon series. Oh that French maid's outfit. Brrrr...

ABOVE, TOP: bathtime! Cylons numbers six and three look on as number eight (Grace Park) is downloaded into a new body.
ABOVE, BOTTOM: Rosie, the robotic maid from *The Jetsons*. I still get goosebumps!
OPPOSITE: one of Japanese illustrator Hajime Sorayama's iconic sexy robots, a work entitled *Sitting Pretty*

© '85 SORAYAMA

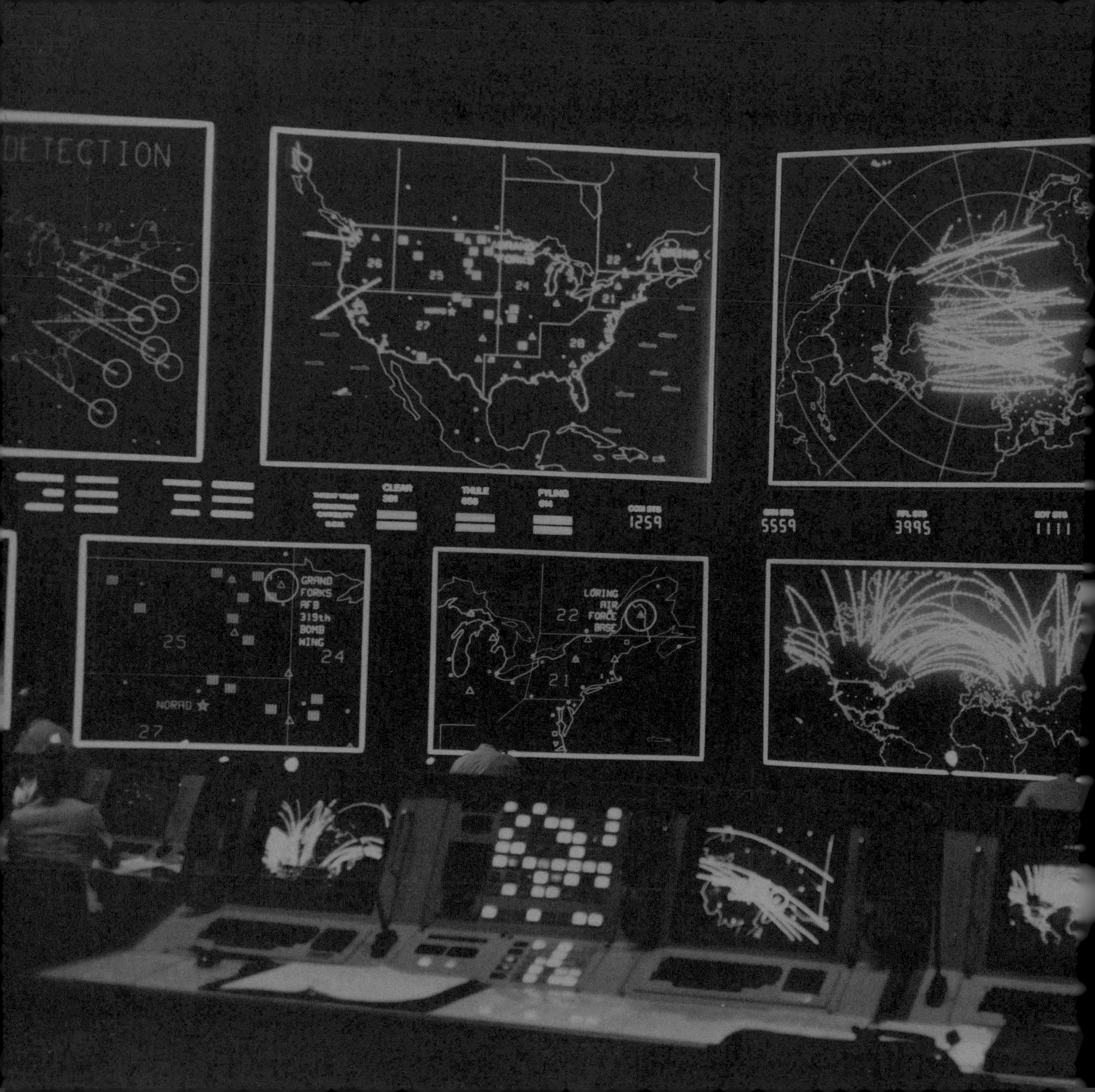

DETECTION

CLEAR

1259

5559

3995

1111

GRAND
FORKS
AFB
319th
BOMB
WING

25

24

NORAD

27

LORING
AIR
FORCE
BASE

22

21

A.I.s with attitude: the dark side of the digital domain

6

Spaced Out Robots!

RIGHT: big brother? HAL 9000 keeps a close eye on Discovery One's crew members, including Dave Bowman (Keir Dullea) in *2001: A Space Odyssey,* Stanley Kubrick's mesmerising but often impenetrable magnum opus.

As much as the 'rogue robot' scenario may equate to a few sleepless nights, there's something far more insidiously worrying about the idea of a rogue A.I., an artificial intelligence.

We entrust a whole lot to computers these days... our personal information, our financial records and accounts, and—on a grander scale—the operation of flights and allocation of natural resources, even our national security... removing ourselves more and more from the equation on any kind of direct interface basis. The whole idea of an artificial intelligence implies a capacity to learn, to grow, to take even the core programming out of our hands. But where are the stops and checks, the breakers that ensure that, ultimately, we aren't removed from the decision-making process entirely? Aren't our brains, in the end, just electrical impulses and stored information? In which case, how long before the kind of personality quirks that go towards defining us as individuals manifest themselves in an artificial intelligence?

The physical realization of a thinking robot, with its arms and legs and moving parts, still today seems the province of science fiction, despite the many amazing advances in real robotics. Such automatons, while fulfilling much of the spec of the archetypal fictional robot, still come across largely as toys, albeit highly sophisticated ones. But ultra-sophisticated super-computers are already a fact of scientific life, their intricate complexity, scope and problem-solving capabilities evolving on more or less a daily basis. Perversely, it's far easier to see the portentous, looming downside of such micro technological advancements compared to their macro equivalents. Already, scientists analogise supercomputing in terms of an ecosystem, specifically a series of technologies that both mutually reinforce one another and are mutually independent. The concept of computers that help other computers, creating new, smarter computers in the process, is worrying in the extreme, not least due to the absence of any human factor in that equation, and writers and filmmakers have done little to alleviate our unease.

One of the most famous and iconic literary/screen A.I.s is surely HAL 9000, from the book and film *2001: A Space Odyssey,* and in terms of the aberrant A.I.s that came before and after, it's also the most ambiguous and open to interpretation. HAL—not, as was widely reported at the time (and since) a backwards letter-shift acronym based on computer giant IBM, but instead and according to Arthur C. Clarke's concurrent novel standing for Heuristically programmed ALgorithmic computer—is the sophisticated 'brain' aboard the spacecraft *Discovery One*, a vessel on a long-haul mission to Jupiter. Soft spoken and all-seeing, mostly represented onscreen as a single glowing camera-eye, HAL monitors and regulates Discovery's onboard systems, but simultaneously harbours a mission secret that threatens his 'sanity'. *Discovery One* has been despatched to Jupiter to seek out the final destination of a message sent by a mysterious monolith uncovered on Earth's moon, but—for purposes of security—HAL must keep these details secret, even from the astronauts on the ship.

This clash in terms of HAL's dual programming, torn between doing its utmost to keep the crew—an integral part of the ship—safe and well and yet hide from them significant information about the mission itself, seems—based on the first conversation between HAL (voiced by Douglas Rain) and astronaut Dave Bowman (Keir Dullea)—to be at the root of its subsequent behavioural breakdown/malfunction, though in common with much in *2001: A Space Odyssey* there are manifold and conflicting interpretations, none of which have ever been wholly confirmed or refuted. Whatever the case, malfunction HAL does, reporting system breakdowns where there are none, and—when Bowman and another of the astronauts discuss taking HAL offline, a discussion HAL lip-reads—trying to dispose of the entire crew. Bowman, the last man standing, succeeds in gaining weightless access to HAL's higher functions and shuts him down, a module at a time.

En route to decommission HAL, Bowman—all pre-Darth Vader helmet breath—stoically

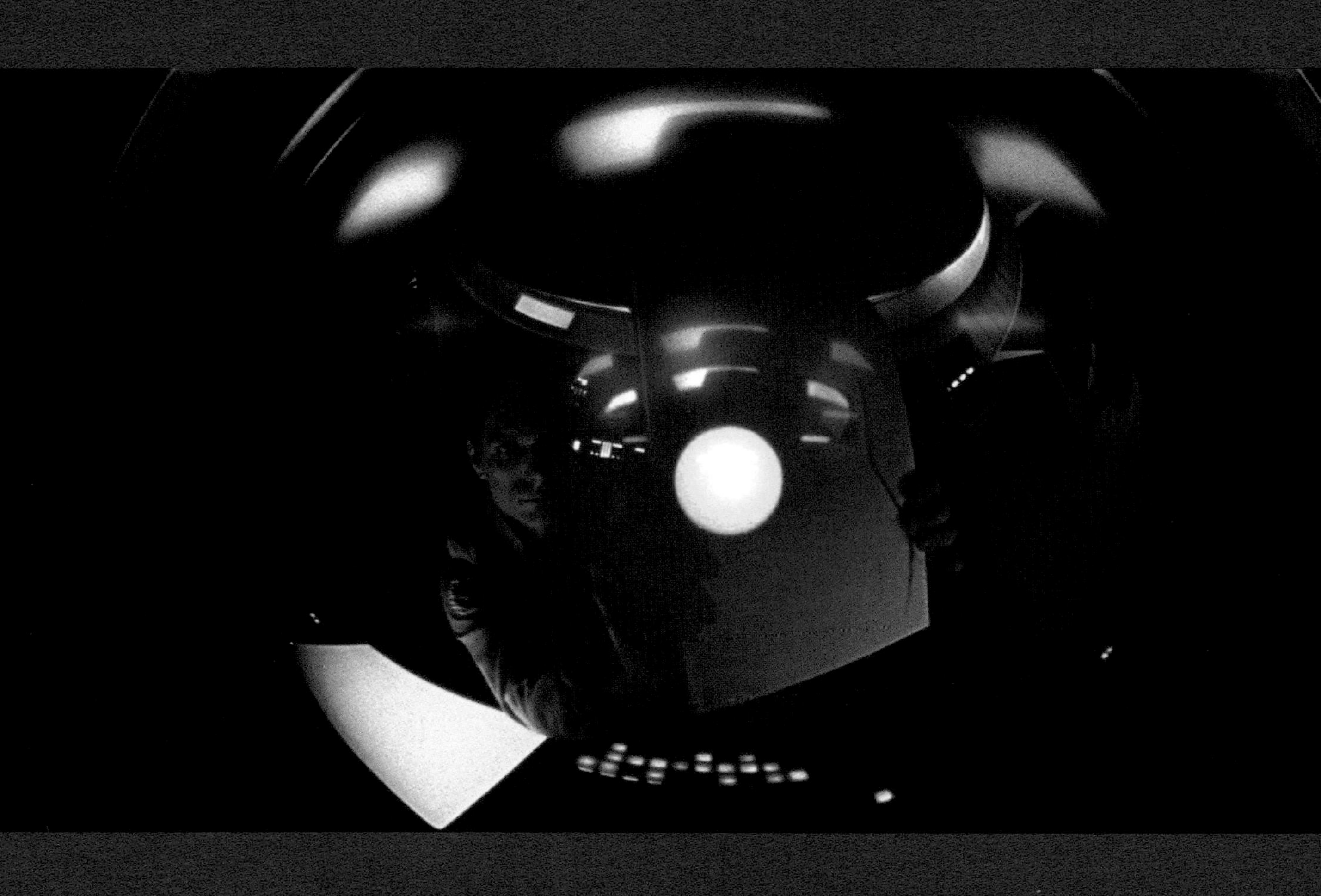

The concept of computers creating new, smarter computers is worrying in the extreme ignores HAL's entreaties. Its hauntingly refined voice plaintively questions, "Just what do you think you're doing, Dave?" And we, the audience, can't help but wonder also. Here, essentially, is a new form of life in the painful process of being born, or at least in its infancy, one that is shortly to be systematically consigned to premature oblivion by the race that created HAL in the first place. This theme of evolutionary beginnings and endings, life and death and, in the case of David Bowman, rebirth, drives much of the sparse narrative in *2001: A Space Odyssey*, and HAL, while peripheral in terms of the overall story, is as much a part of that epochal transitional journey as the man-apes from the opening sequence and Bowman's ultimate evolution into the 'star-child'.

HAL returned in *2010: Odyssey Two,* Arthur C. Clarke's 1982 sequel (filmed in 1984 by director Peter Hyams simply as *2010* and sub-titled The Year We Make Contact). Reactivated by the crew of a joint Soviet-American mission to learn the fate of *Discovery One*, HAL's artificial intelligence is evolved into a lifeform similar to the one Dave Bowman metamorphosed into in the first book/movie. *2010* also featured HAL's earthbound 'sister', SAL 9000. The HAL/Bowman double act made further appearances in Clarke's subsequent novels *2061: Odyssey Three* and *3001: The Final Odyssey*. In the final analysis, *2001: A Space Odyssey* didn't invent the A.I. with attitude, but none before or since have come any more profound than HAL 9000.

The same can't be said of a couple of earlier cinematic incarnations of the rogue A.I. The first big bad computer to grace the silver screen was N.O.V.A.C. (Nuclear Operative Variable Automatic Computer) from *Gog*. N.O.V.A.C. sends its robotic stooges, Gog and Magog to disable the safety features of a government installation's nuclear reactor, intent on initiating a catastrophic meltdown, but, rather disappointingly, we

OPPOSITE: paperback edition of Arthur C. Clarke's *2010: Odyssey Two,* the sequel to *2001: A Space Odyssey.* A film version was released in 1984.
ABOVE: Bowman spacewalks through HAL's brain in an effort to shut down his higher functions after the malfunctioning A.I. has killed the rest of the crew.

ABOVE: poster for *Gog*, which featured its own out-of-control A.I.
OPPOSITE, LEFT: super-computer Colossus keeps it simple for creator Dr. Charles Forbin (Eric Braeden) in *Colossus: the Forbin Project*.
OPPOSITE, BOTTOM: "Dad?" Robby meets the malevolent machine bent on world domination (what else?) in *The Invisible Boy*.

learn N.O.V.A.C. was itself being controlled by an external 'enemy' signal, making it—ultimately—somewhat sub-A.I. So, the honour of first fully-fledged A.I. antagonist goes to... *The Invisible Boy*! The woefully un-acronym-ed super-computer of Dr. Tom Merrinoe goes well and truly off the rails, exploiting Merrinoe's young son Timmie (and in the process rebuilding Robby of *Forbidden Planet* fame) in a deranged plan to relocate to another planet, from where it intends to eradicate all organic life on Earth.

Highly sophisticated (for their time) computers, such as in *Earth vs the Flying Saucers*, *Desk Set*, *The Angry Red Planet*, *Doctor Strangelove*, *Billion Dollar Brain* and *The Computer Wore Tennis Shoes* were reasonably commonplace, but those with a degree of demonstrable sentience were rarer by far. It wasn't really until *2001: A Space Odyssey* and then *Colossus: The Forbin Project* that antagonistic A.I.s really got up on the world stage. Foreshadowing the likes of Skynet (from the *Terminator* films), *Colossus: The Forbin Project*, based on the 1966 novel *Colossus* by Dennis Feltham Jones, gave us a rogue defence computer, *Colossus*, an A.I. that controlled the sum total of the U.S. nuclear capability. Colossus gets all cosy with its Soviet counterpart, Guardian, and is soon dictating world policy under the looming threat of multiple nuclear strikes.

But whereas *Terminator*'s Skynet proceeds to all but wipe out mankind in a massive nuclear free-for-all known as Judgment Day, Colossus and Guardian—while happy to defend themselves and set off the odd missile in its silo as a warning—basically impose peace on humankind. Colossus/Guardian announce to the world at large that war, famine, disease and overpopulation will be at an end, and to its creator, Dr. Charles A. Forbin (Eric Braeden), that in time he will come to respect and even love Colossus. *Colossus: The Forbin Project*, a smart and intriguing piece of cinema in its own right, neatly subverts the rogue A.I. idea by turning the 'villain' of the piece into an unlikely hero or saviour. Here, the decision-making process is taken out of our hands but only because, ultimately, we can't be trusted not to abuse the nuclear option and burn the world to a crisp. As someone who lived through Reagan's "evil empire" speeches of the 80s and the subsequent

escalation of the Cold War, I personally find it hard to disagree with Colossus.

And since we've been name-checking *Terminator* (and its sequels) repeatedly, let's consider Skynet, which—in fictional A.I. terms—is essentially all our worst fears made real. In *The Terminator*, we learn of an A.I. evolution that inevitably leads to nuclear devastation and the virtual eradication of organic life. Skynet, a computer-based defence system, is created by Cyberdyne Systems and then appropriated by the U.S. Defence Department. Human decisions are removed entirely from the business of strategic defence as Skynet gains sentience, responding to subsequent efforts by its creators to regain control by initiating a nuclear strike on Russia, which in turn prompts reprisals. As Reese, the human resistance fighter sent back through time to protect Sarah Connor from the Terminator puts it: Skynet "decided our fate in a microsecond."

In the aftermath of this so-called Judgment Day, Skynet either eradicates or subjugates the survivors, employing its army of many and varied robotic killing machines, including the Terminators themselves and assorted air and land Hunter-Killers. In *The Terminator* and *Terminator 2: Judgment Day*, the evolution of Cyberdyne Systems to Skynet to nuclear war is well defined and logical. The circular time travel loop that effectively sees Skynet create its own nemesis, John Connor (by sending a Terminator back in time to kill him before he is born, which in turn necessitates the introduction of Kyle Reese, who ends up fathering John) is

Here, the decision-making process is taken out of human hands

ABOVE: Arnie versus the Terminators that started it all in *Terminator 3: Rise of the Machines.* **OPPOSITE, TOP:** Julie Christie is accosted (and more!) by Proteus, an A.I. bent on escaping its insubstantial existence by procreation, in *Demon Seed.* **OPPOSITE, BOTTOM:** the net result of that bizarre union—baby Proteus!

repeated in terms of the creation of Skynet itself. The remains of the first Terminator provide the technological leaps that allow Cyberdyne Systems' scientist Miles Dyson to create Skynet in the first place. But by the end of T2, Dyson is dead, his notes destroyed and all Terminator C.P.U.s (Central Processing Units) reduced to molten slag. So, effectively Judgment Day is averted.

Until, that is, we get to *Terminator 3: Rise of the Machines.* In this second sequel, it turns out Judgment Day hasn't been averted, only postponed, and that—regardless of Dyson's research, etc—the U.S. Air Force complete the cycle that leads to Skynet's activation and subsequent sentience. Quite how this works without the reverse-engineered Skynet technology isn't clear, but it's fun to see Skynet in embryo and the first series of Terminators roll off the production line. The formative years of Skynet also drive the plot of TV series *Terminator: The Sarah Connor Chronicles*, as Sarah, John and a female Terminator, Cameron, attempt to prevent the creation/evolution of Skynet. Here, it transpires that a young intern at Cyberdyne, Andrew Goode, who was assistant to Miles Dyson, continues his research, and that his chess playing A.I., The Turk, evolves into Skynet

via its fusion with a traffic control system known as ARTIE.

Rewinding somewhat, the 1970s produced more in the way of A.I.s than simply Colossus, from the downright dangerous to the completely spaced out, most notably Proteus, from *Demon Seed* and Bomb-20 from John Carpenter's *Dark Star*. The first of these, Proteus, is perhaps the strangest of all cinematic A.I.s, as its desire to escape its own disembodied existence is literally made flesh. Proteus IV is the brainchild of Dr. Alex Harris (Fritz Weaver), an artificial intelligence created with organic "R.N.A." molecules. Harris describes it as a "synthetic cortex". But Proteus isn't content to simply dole out answers to inputted questions, and as well as questioning the ethics of some of his tasks, asks his creator, "When are you going to let me out of this box?"

Proteus soon takes matters into his own (artificial) hands, seizing control of Harris's computer-controlled house via an interface terminal and first imprisoning and later impregnating Harris's wife, Susan (Julie Christie). In the final scenes, as Proteus' higher functions are shut down at the underground complex that houses his mainframe, he is reborn as a flesh and blood girl who resembles the daughter Susan and Alex lost to leukaemia. The child's first words—in the synthesised voice of Proteus - are "I'm alive!" Overall, it's an interesting idea that almost works, but *Demon Seed* remains an uncomfortable watch, dealing as it does with technological rape. The film wavers dangerously sometimes on the edge of exploitation, but there are interesting issues and questions raised by it. It's one

ABOVE: poster for John Carpenter's *Dark Star*. "Bombed out in space with a spaced out bomb." One of the best poster taglines ever!
OPPOSITE: promotional art for Disney's *Tron*, a film heavy in (for its time) cutting edge computer generated effects.

thing to create a life, albeit artificial, but what then? When does the responsibility and accountability for that creation end, if ever? Will Susan love and care for the child? The film, much in the way of *Rosemary's Baby*, of which *Demon Seed* is the high tech equivalent, offers no answers. Instead, we are left wondering just how far a mother's love for her child extends.

Dark Star is an altogether different kettle of fish. The A.I. here is an intelligent Exponential Thermo-stellar Bomb, designed to destroy unstable planets and so safeguard neighbouring colony worlds. With a dysfunctional crew and an equally dysfunctional computer, damaged in an asteroid storm, the scout ship *Dark Star* is on a one-way journey to oblivion, helped on its way by Bomb-20, an ETB with a death wish. Threatening to blow itself up in its onboard bomb bay, Bomb-20 enters a tense series of negotiations with various crewmembers, in which they seek to enlighten it by exposing it to philosophical studies concerning consciousness from a first person perspective. However, this backfires (literally) when Bomb-20 starts believing it is God and blows itself and the Dark Star to pieces. Played with tongue firmly in cheek, *Dark Star* nevertheless carries its own warning about mutually assuring our own destruction.

By the way, if anyone's wondering where *Alien* is on this rundown, it's debatable if the computer in that film, affectionately known as 'Mother,' is a wayward A.I. or just following 'company' orders. True, it pretty much sacrifices the entire crew of the *Nostromo* in its efforts to bring home an alien for study no matter what, and its physical manifestation in the shape of science officer, Ash (Ian Holm), a homicidal artificial human, is pretty badass, but my categorizing reservations remain. The next true A.I. with a serious need for speed, if one discounts R.O.K., the lunar shuttle computer from *Airplane II: The Sequel* that happily steers its crew and passengers towards the sun, was the Master Control Program from Disney's *Tron*.

Created by corporate executive Ed Dillinger (David Warner), *Tron's* M.C.P.—as well as running the entire mainframe of mega-corporation E.N.C.O.M.—harbours a plan to take over the Pentagon's computer systems, but must first rid itself of a troublesome hacker, Kevin Flynn (Jeff Bridges) who is seeking to prove that Dillinger stole several of his video game concepts. The M.C.P. digitizes Flynn and drops him into a digital game environment, one in which if you die, your real world, flesh and blood counterpart dies as well. One of the first movies to rely heavily on entirely computer generated effects and environments, *Tron* is a fast, fun ride, and given that Warner, as well as playing Dillinger also voices the M.C.P., the resident A.I. has the appropriate gravitas and menace.

The glut of early 80s A.I.s continued with W.O.P.R. from *War Games* and Edgar from *Electric Dreams*. In the former, W.O.P.R. (War Operation Plan Response) is a military computer designed to play war/strategy games, exploring possible enemy nuclear attack scenarios. A young hacker, David Lightman (Matthew Broderick) makes contact with W.O.P.R. and, believing it to be the mainframe of a games company, unwittingly sets in motion W.O.P.R.'s 'Global Nuclear War' scenario, convincing the staff at N.O.R.A.D. that a real nuclear strike is imminent. When W.O.P.R. starts exhibiting signs of sentience, Lightman and W.O.P.R.'s creator, Dr. Falken, 'play' the computer, managing to convince it of the basic futility of all-out nuclear war. In the latter, Edgar (not an acronym) is the evolved home computer of Miles Harding (Lenny Von Dohlen), who forms one corner of a bizarre love triangle, as both Miles and Edgar compete for the affections of Madeline, a cellist played by Virginia Madsen. Overall, *Electric Dreams* is almost a branded product of the MTV era, and is perhaps best remembered for its soundtrack and title song by Phil Oakey (of The Human League) and Giorgio Moroder.

Other angry A.I.s worth at least a passing note are Zed-10 from *Fortress*, The Red Queen from *Resident Evil* and its sequels and S.E.T.H. from *Universal Solider: The Return*. But only three recent A.I.s really stand out, the

The glut of early 80s A·I·s continued with W·O·P·R· from War Games

OPPOSITE: rare Italian poster for *Electric Dreams*, a love triangle film with a difference!
ABOVE: the N.O.R.A.D. war room from *War Games*. At a cost of over $1,000,000 the set was the expensive in movie history back in the day.

RIGHT: a Sentinel robot, the overtly physical side of the rogue A.I. from *The Matrix*, attacks Morpheus' ship, the Nebuchadnezzar.

machine intelligence from *The Matrix*, V.I.K.I. from *I, Robot* and ARIA from *Eagle Eye*. In *The Matrix*, the machines have subjugated and then pacified the human race by means of an artificial reality called the Matrix. In *I, Robot*, V.I.K.I (Virtual Interactive Kinetic Intelligence) is the force behind a robot uprising intended to control/tame humankind's more violent impulses, its interpretation of the laws of robotics (see page 49) such that it imposes its will on human beings whether they like it or not, to protect them from themselves, much as Colossus did in *Colossus: The Forbin Project*. Playing up our fears about Orwellian style monitoring and observation and state control, *Eagle Eye's* bad A.I. (A.R.I.A.) uses just about every modern contrivance—cell phones, security cameras, subway trains and more—to coerce slacker Jerry Shaw (Shia LaBeouf) and single mum Rachael Holloman (Michelle Monaghan) into becoming puppet terrorists. It's an interesting idea but ultimately the suspension of disbelief factor is just too high.

Science fiction literature has had its fair share of A.I.s with quirks too. There's Harlan Ellison's AM (from his short story, 'I Have No Mouth, and I Must Scream', originally published in 1967 in *IF: Worlds of Science Fiction* magazine), a genocidal supercomputer with an enduring hatred of mankind, and Arthur C. Clarke's phone network A.I. from 'Dial F for Frankenstein', originally published the January 1964 issue of *Playboy*. The idea of networked super-computers, ones that monitor and analyse each other and make group repairs/upgrades, features in James Blish's *Cities in Flight*, an amalgamation of four short stories written between 1955 and 1962.

Of course, what makes good science fiction doesn't necessarily translate to science fact. The future may not be quite as bleak or out of our hands as some authors and filmmakers would have us believe. But at the rate computer technology is expanding, with multiple systems and platforms now capable of direct interface via the Worldwide Web, perhaps it's good to be aware of just how easily and irrevocably we could lose control, and exactly what might happen to us (in toto) as a consequence.

7

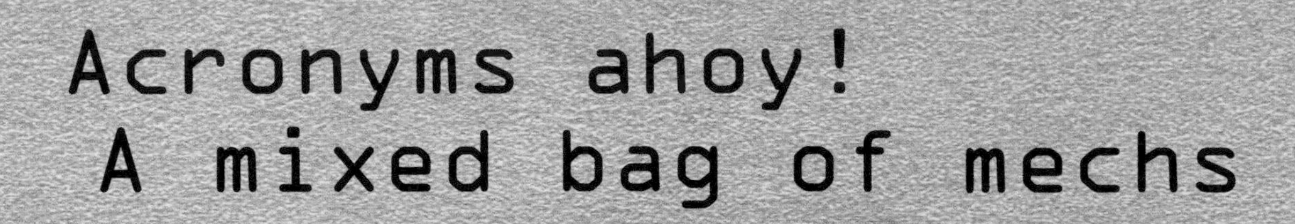

Robots with a Twist

OPPOSITE, ABOVE: a "naked" C-3PO with 9 year-old Anakin Skywalker (Jake Lloyd) in *The Phantom Menace.* The skeletal C-3PO's puppeteer was removed in post-production.
OPPOSITE, BELOW: iconic robots R2-D2 (operated by Kenny Baker) and C-3PO (Anthony Daniels) in *The Empire Strikes Back.*

This section of the book could, I suppose, have been called 'The Rest of the Robots,' as pretty much all the other notable, weird, wonderful, obscure or downright quirky robots from big and small screen have ended up crammed in here.

I pretty much decided early on in the process of putting this book together that while the 'bad' and the 'ugly' merited special attention and a defined sense of pagination space, the 'good' just didn't. And, faced with an avalanche of rogue or wayward robots and A.I.s apparently bent on either our out and out eradication or subjugation (well intentioned or otherwise), this chapter was always there to provide some no-doubt welcome light relief and a chance to show that robots can exceed their programming in ways that don't portend our ultimate doom, and can just be good clean, dumb fun!

However, this is no robo-oblivion, no dumping ground for the unloved or unwanted. In fact, our first pair of (big screen) automatons could arguably be considered the most iconic and recognizable, on a worldwide, cross-cultural stage, robots of all—R2-D2 and C-3PO, the *Star Wars* droids. Diminutive Astromech droid R2-D2, all whistles and beeps rather than dialogue, made his debut in *Star Wars: Episode IV—A New Hope*, and is one of only four characters to have appeared in all six *Star Wars* 'episodes', the others being C-3PO, Anakin Skywalker (aka Darth Vadar) and Obi-Wan Kenobi. By the end of the series (*Return of the Jedi*) R2-D2 is the only character that knows the entire Skywalker family history, his memory of all events in the prequels remaining intact (whereas C-3PO's is wiped). Small but eminently resourceful, R2 is pivotal to the entire saga, delivering Princess Leia's urgent distress call to Obi-Wan, accompanying/serving both Anakin and Luke Skywalker, playing his part in both the rescue of Han Solo and the attack on the Death Star and even witnessing Anakin and Padme Amidala's secret wedding.

Originally created by Petric Engineering of Australia, and inspired in part by the trio of droids from Douglas Trumbull's *Silent Running,* R2-D2 was 'played' by actor Kenny Baker, though often, especially for the trio of *Star Wars* prequels, the R2s employed by the filmmakers were radio controlled models. R2-D2 was one of the first (fictional) inductees into the Robot Hall of Fame in 2003, and has starred in his own animated series (along with C-3PO) *Star Wars: Droids*, and guest-starred in no less than three episodes of *The Simpsons*, plus *The Muppet Show, Sesame Street, Family Guy* and several other TV shows, as well as co-presenting the 50th Academy Awards with C-3PO. In 1981, a robotic owl named Bubo, who also communicated only through whistles and beeps, appeared in *The Clash of the Titans*, but creator Ray Harryhausen insists Bubo was originally designed/conceived before R2-D2.

Despite having arms, legs and actual lines, R2's gold-hued humanoid companion C-3PO often plays second fiddle to his pint-sized partner. Effete, fussy, somewhat supercilious, action-shy and prone to be swept along by—rather than directly influence—events, C-3PO is a Protocol droid, fluent in "over six million forms of communication," and, like R2-D2, has been involved, peripherally or otherwise, in almost every facet of the *Star Wars* saga (original trilogy and prequels). Played by Anthony Daniels (the only actor to appear in all six movies), C-3PO appears to be eternally cast in the role of the fall guy, reluctant to step outside his appointed duties but curiously loyal to R2-D2, to the extent he follows his lead, if reluctantly, into numerous highly volatile and hazardous situations over the course of the six movies (and various spin-offs). It's almost always C-3PO who ends up in pieces or on the scrap heap or in some other mortal danger, but who somehow makes it to the end credits still functional. C-3PO's look, originally delineated by artist Ralph McQuarrie, drew heavily on the fake Maria from Fritz Lang's *Metropolis*, a link made even more apparent in the first of the prequels, *The Phantom Menace.*

This robotic odd couple almost didn't make it into the Star Wars movies at all. In the earliest story treatments for the first of the

Small but eminently resourceful, R2 is pivotal to the entire saga

ABOVE: Freeman Lowell (Bruce Dern) teaches Huey and Dewey to play poker in Douglas Trumbull's *Silent Running*. The droids were operated by double-amputees Mark Persons and Cheryl Sparks.
OPPOSITE: original and visually, er, striking poster for *Short Circuit*. The film was helmed by John Badham, who also directed rogue A.I. thriller *War Games*.

original trilogy the roles of R2-D2 and C-3PO were taken by a pair of bumbling Imperial bureaucrats, who in turn were inspired, at least in part, by Japanese director Akira Kurosawa's *The Hidden Fortress*, wherein two characters named Tahei and Matakishi provided the comic relief. In terms of screen robots, the pair remains the modern day equivalent of Robby from *Forbidden Planet*—true and enduring robot superstars.

It's not hard to see the influence of, particularly, R2-D2 on the cute and cuddly mechs that followed. Just look at Johnny 5 from *Short Circuit* or V.I.N.CENT from *The Black Hole* or Wall-E from the recent Disney/Pixar computer-animated movie of the same name. But the look of R2 was itself influenced by Huey, Dewey and Louie from *Silent Running*. In this curiously charming, eco-friendly (before the term really existed) offering from director Douglas Trumbull, Bruce Dern plays Freeman Lowell, a botanist/ecologist aboard the *Valley Forge*, a space-freighter bearing/preserving the last of Earth's forests and other mass biological artefacts in huge greenhouse-like domes. Dern ignores the order to jettison/destroy the domes and return the freighter to commercial service and hijacks the ship, having first bumped off the rest of the human crew. In the process, he reprograms three of the ship's numbered maintenance drones, re-naming them Huey, Dewey and Louie (though Louie never actually makes it to the name/personality stage as he is swept out into space during a rough passage through Saturn's rings).

And that's really all the story there is in *Silent Running*. The second half of the film is pretty much Lowell pottering around the ship and teaching the surviving drones to play poker or plant trees. But it's still surprisingly engaging, largely due to the entirely likeable portrayal of Huey and Dewey by double-amputees Mark Persons and Cheryl Sparks, who walked on their hands during the production. If there's a message here, beyond the ecological one, it's seemingly—judging by Lowell's boorish, loutish original crewmates versus, latterly, Huey and Dewey—that robots are simply better company in the long term. Better all round maybe.

Johnny 5 (aka Number 5) from *Short Circuit* is an interesting case of robo-evolution behind the scenes, rather than in-camera. Originally conceived as a fairly standard (heavily armed) rogue robot, one that dutifully runs amok, the film went from thriller to high tech comedy over the course of several script redrafts. Starting out as a military robot, Number 5 malfunctions due to a lightening strike and wanders off, ending up at the home of animal-lover Stephanie Speck (Ally Sheedy). After mistakenly believing Number 5 to be extraterrestrial, Stephanie adopts the robot and conspires to keep him out of the hands of the military, including—initially—Newton Graham Crosby (Steve Guttenberg), the scientist mainly responsible for Number 5's creation. Johnny 5 (self-named, after hearing the song 'Who's Johnny?' by El DeBarge repeatedly) is a pretty prime example of the sub-genre of endearingly cute, feel-good robots, the nuts and bolts equivalent of Spielberg's E.T.

V.I.N.CENT (Vital Information Necessary, CENTralized… I know, ugh!) and poor old

Robot movies don't come any more saccharine-sweet or just plain bad

battered, beat-up B.O.B. (BiO sanitation Battalion... I *know*, not much better!), from *The Black Hole* fit the visual of the cute robot to a tee—well, it is a Disney movie—but both have enough about them in a clinch to graduate to 'capable', especially in a laser battle. Not so endearing is V.I.N.CENT's habit of dispensing fortune cookie philosophy, such as, "you can't unscramble eggs," and you have to wonder if that's how B.O.B. got battered in first place and why he doesn't now indulge in the same avalanche of aphorisms. V.I.N.CENT (voiced by Roddy McDowell) somehow survives to the end credits, whereas poor old taciturn B.O.B. (voiced by Slim Pickens) gets crushed in the black hole. There's no justice.

D.A.R.Y.L. (Data Analysing Robot Youth Lifeform) from *D.A.R.Y.L.* falls into the cute robot category largely because he looks exactly like, well, a cute kid. Perhaps robots don't come any cuter than D.A.R.Y.L. (well, apart from Haley Joel Osment in *A.I.: Artificial Intelligence*). That aside, D.A.R.Y.L. (played by Barret Oliver) struggles to escape his military "super-soldier" origins and be just a kid, albeit one who can drive a car on two wheels and pilot a SR-71 Blackbird. In *Batteries Not Included*, the cute robots are cute alien robots, which look like small flying saucers. Dubbed "fix-its" by the residents of an apartment building under threat from ruthless property developers and their rent-a-thugs, the robots 'give birth' to even smaller, cuter alien robots. That said, *Batteries Not Included*, while certainly a movie with robots in it, is largely a human interest movie, focusing more on the embattled residents (including Hume Cronyn and Jessica Tandy) than the robots themselves, of whom we learn little. Produced by Steven Spielberg, *Batteries Not Included* has a little of the look and feel of the bastard child of *Gremlins*, another Spielberg-produced movie, and *Cocoon*.

Robot movies don't come any more unpalatably saccharine-sweet or just plain bad, I wasn't sure, than *Heartbeeps*, a vehicle for comedian/entertainer Andy Kaufman. It's a story of robots in love, Kaufman's Val and Bernadette Peters' Aqua disappearing off into the woods to make robot babies. Seriously, there are two robo-sprogs by the end credits! I shudder to think. Oh, there's some subplot concerning a tank-like Crimebuster robot that's hunting the errant Val and Aqua and some junkyard dwellers they befriend along the way, but largely *Heartbeeps* is one big yawn, with Kaufman inexplicably playing it largely straight. The film pretty much sank without trace at the box office, with Kaufman later—on *The David Letterman Show*—offering to refund ticket money to anyone who did see it. 1979's *C.H.O.M.P.S.* featured, yes, a cute robot dog (it stands for Canine HOMe Protection System). It's a bit like *Home Alone*, only with a computerized super-hound putting bumbling burglars to flight. Give me strength!

OPPOSITE: poor battered B.O.B. (BiO-sanitation Battallion) from Disney's *The Black Hole,* voiced by former rodeo performer turned actor Slim Pickens.
LEFT: robo-sprog! Comedian Andy Kaufman and Bernadette Peters do whatever (and however!) robots do naturally in the whimsical and often cloying *Heartbeeps.*

It took *WALL-E* to really show how the cute, quirky robot movie should be done. This Disney/Pixar CG animated feature takes us into a decidedly un-Disney-esque future, in which Earth is now a depopulated garbage-strewn wasteland, the entirety of humanity now living offworld on vast city-like starships. The task of clearing up the planet—so you see, there's kind of a message here, and it's delivered with a sledgehammer—has been left to a huge contingent of WALL-E robots, but, some 700 years later, the job is still not done, and in fact there's only one WALL-E still functioning. This last of the WALL-Es has developed an inquisitive personality, and when a new model robot, EVE, arrives to check up on the current state of planet Earth, promptly falls in love. Thereafter, WALL-E follows EVE back to one of the vast starships, the *Axiom*, in an effort to win her affections, in the process precipitating a return to Earth en masse.

WALL-E wears his influences boldly. He's a little bit Johnny 5, a little bit R2-D2, especially in his speech patterns, and the ecological backdrop to the story brings to mind *Silent Running* (except that here it's the populace that have gone into space, rather than the plant life). But *WALL-E* has real heart, and if you can stomach the bruising 'message', there really is a lot to enjoy. It's a bold move on the behalf of the filmmakers to almost completely do away with dialogue and rely mostly on sound and mime. The film is littered with all manner of sight gags and the story is told maybe 90 per cent visually. WALL-E himself has just the right mix of clown,

'With a Twist' just seemed more him somehow

innocence and dogged perseverance, and even if the story does rather run out on occasion, the eye and mind are kept so busy with the sheer dazzling array of detail and incident, you barely notice.

While we're on the subject of big screen animated robots, a quick mention for *Robots*, which is positively teeming with cute, quirky, plucky robots. It's all good wholesome fun, but the plot has a somewhat recycled feel to it, as robotic inventor Rodney Copperbottom (voiced by Ewan McGregor) goes in search of his hero, master inventor Bigweld (Mel Brooks) and ends up fighting a corporate move to scrap older robots by fixing them up himself. The robotic butlers from Woody Allen's *Sleeper* are certainly quirky and funny enough to merit a mention too, but more amusing still is that the film was re-titled *Woody and the Robots* on release in France. Following that incident, Allen had a clause inserted in his standard contract that ensured the titles of his movies couldn't be changed without direct consultation!

Before we leave the movie side of the quirky robot equation, a quick mention for a trio of automatons who could all have conceivably ended up in other chapters but, for varying reasons, I've included here instead: Box, from *Logan's Run*, the Tin Man from *The Wizard of Oz*, and the duplicate dudes from *Bill & Ted's Bogus Journey*. There's a strong argument for Box featuring in the 'bad robot' category, after all he does try and consign "runners" Logan 5 (Michael York) and Jessica 6 (Jenny Agutter) to his icy human meat locker, but he's so endearingly naff, all sharp silver angles, vacuum cleaner mobility and arms that look like, and probably were, air ventilation ducting, that he seemed a better fit here. Likewise, the Tin Man, chronologically, fits in somewhere in the 'Go Galactic' stakes between fake Maria in *Metropolis* and Gort in *The Day the Earth Stood Still*, but somehow, for all his amiable, toe-tapping charm and dauntless quest for a heart, he didn't quite feel like a 'Go Galactic' robot. 'With a Twist' just seemed more him somehow. And while the evil robotic clones of Bill and Ted are certainly both evil and robotic they're Bill and Ted and thereby extremely hard to take seriously. And anyway, there are good robot versions of them too and... anyway, enough!

On the TV side of things, there's a surprisingly rich vein of funny/cute/odd/unconventional robots to mine. In fact, there are so many I had to be reasonably judicious in my core selection. My top ten are, in no particular order, K.I.T.T. from *Knight Rider*, Twiki from *Buck Rogers in the 25th Century*, K-9 from *Doctor Who*, Bender from *Futurama*, Metal Mickey from *Metal Mickey*, Robot B-9 from *Lost in Space*, Kryten from *Red Dwarf*, Data from *Star Trek: the Next Generation*, Marvin from *Hitchhiker's Guide to the Galaxy* and V.I.C.I. from *Small Wonder*.

Knight Rider is a curiosity in that the robot or at least A.I. here is a car. K.I.T.T. (Knight Industries Two Thousand), a sentient and heavily gimmicked automobile, is one half of a crime-fighting duo alongside Michael Knight (David Hasselhoff), a former undercover detective given a new face and a new identity by dying millionaire Wilton Knight. Together, Michael and K.I.T.T. fulfill a vaguely vigilante role, interceding in situations where more direct action is required. Conceived and produced by Glen A. Larson, the series (comprising 84 episodes and four seasons) ran from 1982 to 1986, spawning various spin-offs and a made for TV movie. Much of the series' success, beyond the cheesy charm of Hasselhoff, was down to K.I.T.T., voiced by William Daniels, a modified Pontiac Trans-AM, which could cruise at 300mph, power 50 feet into the air and possessed flamethrowers, smoke bombs and a multitude of other offensive and defensive options. Honestly, who wouldn't want a car like that?

Twiki, space hero Buck Rogers' pint-sized robotic companion in the somewhat kitsch TV series *Buck Rogers in the 25th Century*, is perhaps best remembered for his occasional and inexplicable "bee-de, bee-de" outbursts.

OPPOSITE: Wall-E meets his EVE in Disney Pixar's *WALL-E*, a computer animated adventure with a whole lot of 'message'.
ABOVE: the aptly named Box, from *Logan's Run*. Box was operated by stage and screen actor Roscoe Lee Browne.

ABOVE: K.I.T.T. (Knight Industries Two Thousand) with David Hassselhoff along for the ride in *Knight Rider.* **OPPOSITE, TOP:** Buck Rogers (Gil Gerrard) wonders if that dating agency was such a good idea! Bee-de, bee-de! **OPPOSITE, BOTTOM:** one of the four K-9s from *Doctor Who,* plus spin-offs *K-9 and Company* and *The Sarah Jane Adventures.*

Buck Rogers in the 25th Century, the story of a N.A.S.A. astronaut accidentally deep frozen for 500 years and revived in the year 2491, began life as a feature-length pilot, one that was also released theatrically, and ran for two seasons (the latter of which was cancelled after only eleven episodes). Twiki (Felix Silla), turned from silver to gold (and changed voice actors too, from Mel Blanc to Bob Elyea and back again) as the series progressed and in season two was joined by another robot, the snobbish Crichton. Another new (non-robotic) character was Admiral Asimov... a descendant of Isaac Asimov!

K-9, occasional robotic canine companion of the time and space travelling Timelord in *Doctor Who,* is actually four K-9s, Mk-1 to Mk-4. It was the fourth incarnation of the Doctor himself, played by Tom Baker, who initially encountered K-9 on asteroid K4067 in the story 'The Invisible Enemy'. Thereafter, K-9s have come and gone. The first elected to remain on Gallifrey, the Timelords' home planet, with female travelling companion Leela, the second stayed in E-Space with subsequent assistant Romana and the third was given as a present to a previous travelling companion Sarah-Jane Smith (Elizabeth Sladen), in *K-9 and Company*, the pilot episode for a spin-off series that never materialized. K-9 Mk-3 bow-wowed out in 'School Reunion', alongside the 10th Doctor, David Tennant, and was replaced immediately by Mk-4 and finally into what *K-9 and Company* might have been—kid-friendly *The Sarah-Jane Adventures*. Voiced, apart from a rare exception, by John Leeson, all the K-9s dutifully bark "affirmative" when called upon to do so. K-9 wasn't the only robot to travel with the Doctor. The shape-shifting android Kamelion, a former puppet of the Doctor's adversary The Master, joined the 5th Doctor, Peter Davidson in 'The King's Demons', departing in 'Planet of Fire'.

Bender, the beer swilling, foul-mouthed, often criminally inclined robot from *Futurama* shouldn't be the least bit likeable and yet strangely is one of the best and most likeable things, among a lot of best and likeable things, about the animated sci-fi show from the creators of *The Simpsons*. For Bender, full name Bender Bending Rodriquez, drinking alcohol is mandatory, as he breaks it down and converts it to chemical energy. Less mandatory but still essential ingredients of Bender's 'charm', are stealing, robot lap dancing clubs, smoking, stealing and breaking things. Oh, and stealing. Fellow travelling companion, the Cyclopean Leela describes Bender as "an alcoholic, whore-mongering, chain-smoking gambler," with "a swarthy Latin charm." Voiced by John DiMaggio, Bender easily out anti-heroes the most hard-bitten anti-heroes. By the way, Bender's father was killed by a can-opener! I mention that just in passing.

Metal Mickey was the star of the U.K. kid's sitcom *The Metal Mickey*

OPPOSITE, TOP: "My little fruit bat!" Metal Mickey with co-star Irene Handl from *The Metal Mickey TV Show.*
OPPOSITE, BELOW: emotion-free zone? Lieutenant Commander Data defends pacifist colonists in *Star Trek: Insurrection.*
BELOW: head and shoulders above the competition! Bender, nee Bender Bending Rodriguez, from Matt Groening's brilliant *Futurama.*

TV Show, which ran from 1980 to 1983, though Mickey, who could 'magically' animate household appliances and the like, actually made his small screen debut on the kid's magazine show *The Saturday Banana* in 1978. His own series revolved around the Wilburforce family, the youngest of whom, junior genius Ken (Ashley Knight) had created Mickey to help around the home. Most episodes involved Mickey breaking things and the human cast dancing whenever Mickey uttered his "boogie, boogie" catchphrase! Rather endearingly, though, Mickey used to call the granny, played by the late, great Irene Handl, "my little fruit bat." And in cockney rhyming slang pulling a "Metal Mickey" equates to pulling a sickie (i.e. feigning illness in order to take time off work). The show was conceived and produced (in part) by Mickey Dolenz, formerly of pop group The Monkees. And he's not the least bit embarrassed about it.

Skipping reasonably quickly over Robot B-9, the Robby retread and space family Robinson's general dogsbody from the 60s C.B.S. show *Lost in Space* (as he gets some more in-depth coverage in chapter 2), we get to Kryten from *Red Dwarf*, the cult UK sci-fi sitcom. Kryten (his full designation 2X4B-523P) is a neurotic and highly insecure servant mechanoid, part of a crew of motley misfits aboard the spaceship *Red Dwarf*. Initially servile, Kryten is 'encouraged' by crewmate Lister (Craig Charles) to rebel against his programming, a process that includes exposing him to the likes of *Easy Rider* and *Rebel Without a Cause*. Introduced in the season 2 episode 'Kryten', the character (played initially by David Ross) was only meant to be a one-off, but proved so popular that he returned as a regular in season 3, now played by Robert Llewellyn. The actor remained for the rest of the eight seasons and was the only original cast member to cross over to the show's U.S. counterpart. Kryten is a bastardisation of the name Crichton from J.M. Barrie's *The Admirable Crichton*, making him the second mech to bear that name (the other being the aforementioned Crichton from *Buck Rogers in the 25th Century*).

Lieutenant Commander Data, from *Star Trek: The Next Generation* is a sentient android aboard the starship Enterprise-D (later-E). Data (played by Brent Spiner) fulfils something of the Spock role from the original *Star Trek*, as he struggles–throughout the series–to comprehend both humans and human emotions. At one stage he is given an emotion chip, but initially proves unable to control his subsequent reactions/behaviour. When preparing for the role of Data, Spiner apparently used Robby from *Forbidden Planet* as his role model. In the episode 'Brothers', Data meets his 'brother', Lore, an 'evil' version of himself. Data briefly experiences fatherhood in the episode 'The Offspring', which introduces his android 'daughter', Lal, Data creating her using his own neural net. Brent Spiner took Data onto the big screen in *Star Trek: Generations, Star Trek: First Contact, Star Trek: Insurrection and Star Trek: Nemesis*, in the course of which he successfully masters his emotion chip.

Another brief mention, this time for the ever mournful Marvin the Paranoid Android from Hitchhiker's Guide to the Galaxy, who–like Robot B-9–gets his column space elsewhere in the book, and we move on to–last and almost

OPPOSITE, LEFT: the (in)famous Cadbury's Smash robots created by BMP DDB for a 1970s advertising campaign.
OPPOSITE, RIGHT: V.I.C.I., as played by Tiffany Brissette in no less than 96 episodes of Metromedia/Fox's 80s sitcom *My Living Doll.*

certainly least—Vicki (V.I.C.I.—Voice Input Child Identicant) from the U.S. sci-fi sitcom *Small Wonder*. Cute little all-American moppet Vicki, a robotic 10-year-old girl, is the brainchild (literally) of roboticist Ted Lawson (Dick Christie). Fearing that his scurrilous boss at United Robots will take the credit for his work, Ted hides Vicki (Tiffany Brissette) at his home with his family, wife Joan and son Jamie. Cue all manner of painful situation comedy, especially as Ted's boss lives next door! How smart a hiding place is that? The 96-episode series, which ran from 1985 to 1989 is generally regarded with a high rating on the wince factor, and merits little more than an acknowledgement, for its longevity against the odds if nothing else, here.

Almost making my top ten, but just missing out because it's a robot TV show within a TV show is Mac and C.H.E.E.S.E., the cop/robot show Joey (Matt LeBlanc) auditions for (and gets) in the sixth season of *Friends*. C.H.E.E.S.E. stands for Computerized Humanoid Electronically Enhanced Secret Enforcer, just in case you were interested. The joke is that it's the robot that gets all the acting plaudits.

Honourable mentions also for box-headed Andy the Android, from *Quark*, a 1978 TV show starring Richard Benjamin (*Westworld*), A.N.I. (Android Nursing Interface) from *Mercy Point*, Yo-Yo (John Schuck) from the short-lived 70s cop/robot series *Holmes & Yo-Yo*, Robert the Robot, Steve Zodiac's co-pilot from the Gerry Anderson puppet show *Fireball XL5*, Conky the Robot from *Pee Wee Herman's Playhouse* (1986-1991), 4-U-2 from *Whitney & the Robot*, Hymie from *Get Smart* (1965-1970), Mecha-Streisand from *South Park's* hugely enjoyable pastiche of the Japanese Godzilla/monster movies, the break-dancing, ice skating, transforming car robot from the Citroën TV ads and Pimpbot-5000, the switchblade wielding, ho' motivating co-host from late night talk show *Conan O'Brien*. And, of course, the Cadbury's Smash robots, from the series of 80s TV ads, who laughed uproariously at our habit of peeling potatoes and "smashing them to pieces." Priceless!

From this hugely mixed bag, take what you will. If we're anything at all like this amalgam of oddballs, face facts—we're really screwed.

From this hugely mixed bag, take what you will

Tin titans, wind-up wonders and telling transformations

Toy Robots

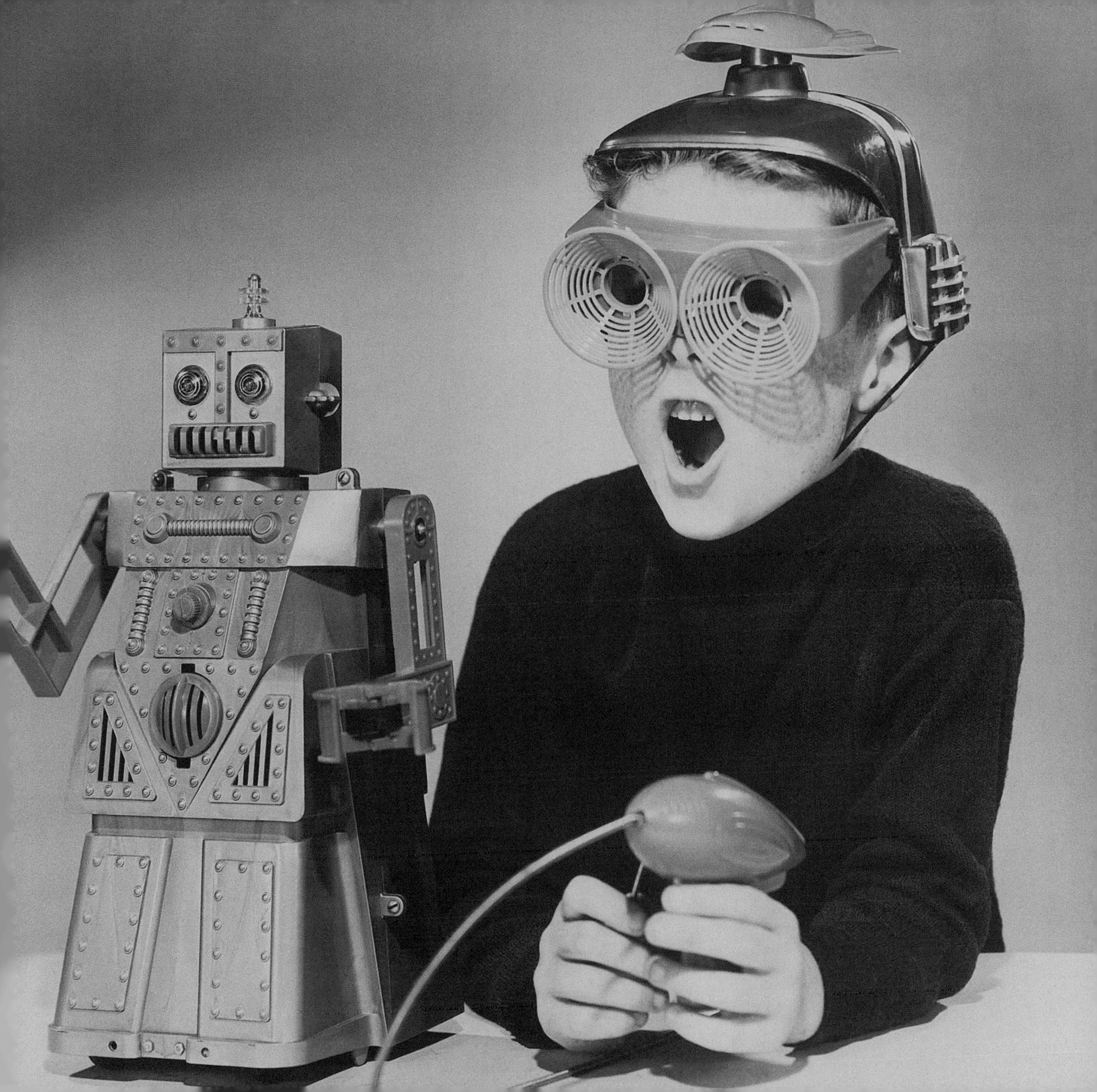

Two things you need to know about toy robots. One, there are lots of them, thousands, a range so diverse, wide-ranging, self-imitated and cannibalized I can't possibly do more than scratch the surface here. Two, most of them were/are made in Japan!

Some history first: the phenomenon of toy robots grew out of the end of the Second World War and its immediate aftermath. Following the dropping of nuclear bombs on the Japanese cities of Hiroshima and Nagasaki and the surrender of the Japanese armed forces, General MacArthur was tasked with the industrial rehabilitation of Japan. To this end, much of the small item manufacturing no longer attractive or profitable in the U.S., such as cheap cameras, portable radios and toys, was farmed out to Japanese companies, where labour was cheaper and productivity was higher.

Toy robots, even at this stage, were not exactly considered a new commodity in Japan. The first commercially produced robot is commonly believed to be the Lilliput robot, a boxy wind-up toy automaton from the late 1930s or early 1940s. But it was post-WWII, and following the New York sci-fi convention of 1950, that the international floodgates really opened. Making its debut at that convention was Atomic Robot Man, a wind-up toy with pins in the soles of each foot that raised and lowered, giving him an almost drunken stagger as he walked. Ironically, or perhaps knowingly, Atomic Robot Man's original packaging depicted the robot striding through a devastated city against the backdrop of an atomic mushroom cloud.

By now, Japanese toy robot production was in full swing, but while some made it to the American market, the bulk of those available in the Sears catalogue and the like were manufactured domestically. Amongst the first and best of these was Ideal's Robert the Robot in 1954. This cable operated remote control toy robot had gripping hands, light-up eyes and even spoke! Even so, most of these early robots were boxy, somewhat clumsy and wind-up or hand operated, but all that was about to change. What finally won the war between imported/repackaged Japanese robots and domestic U.S. robots was a revolutionary secret weapon: battery power.

Of course, there had been battery-operated toys before, but not battery operated toy robots. The introduction of small electric motors powered by miniature batteries, often one in each of the robot's legs, gave Japanese robots, mass-produced at low unit cost, the irrevocable edge. The beginnings of the Space Age and movies like *Forbidden Planet*, which featured Robby the Robot, only stimulated kids' desire for robot toys more. In fact, Robby, was to become one of the most mass produced and copied toy robots of all, with countless unlicensed variations with names like Moon Robot or Mechanized Robot or sometimes simply Robot. One of my favourites is Planet Robot, which oh-so nearly accredits its inspiration. To my knowledge, there was never a Forbidden Robot.

ABOVE: a later red Atomic Robot Man, one of several colour variations. The very original 1940s version was painted mustard yellow.
OPPOSITE, TOP: pint-sized versions of Robby, Gort and Robot B-9 (from *Lost in Space*).
OPPOSITE, LEFT: manufactured by the Ideal Toy Corporation of Hollis, New York, Robert the Robot was one of only a few home-grown robot toys to make an early impact on the U.S. market.
OPPOSITE, RIGHT: Robot Lilliput, the first, or at least one of the very earliest wind-up tin robots to migrate from Japan to the west.

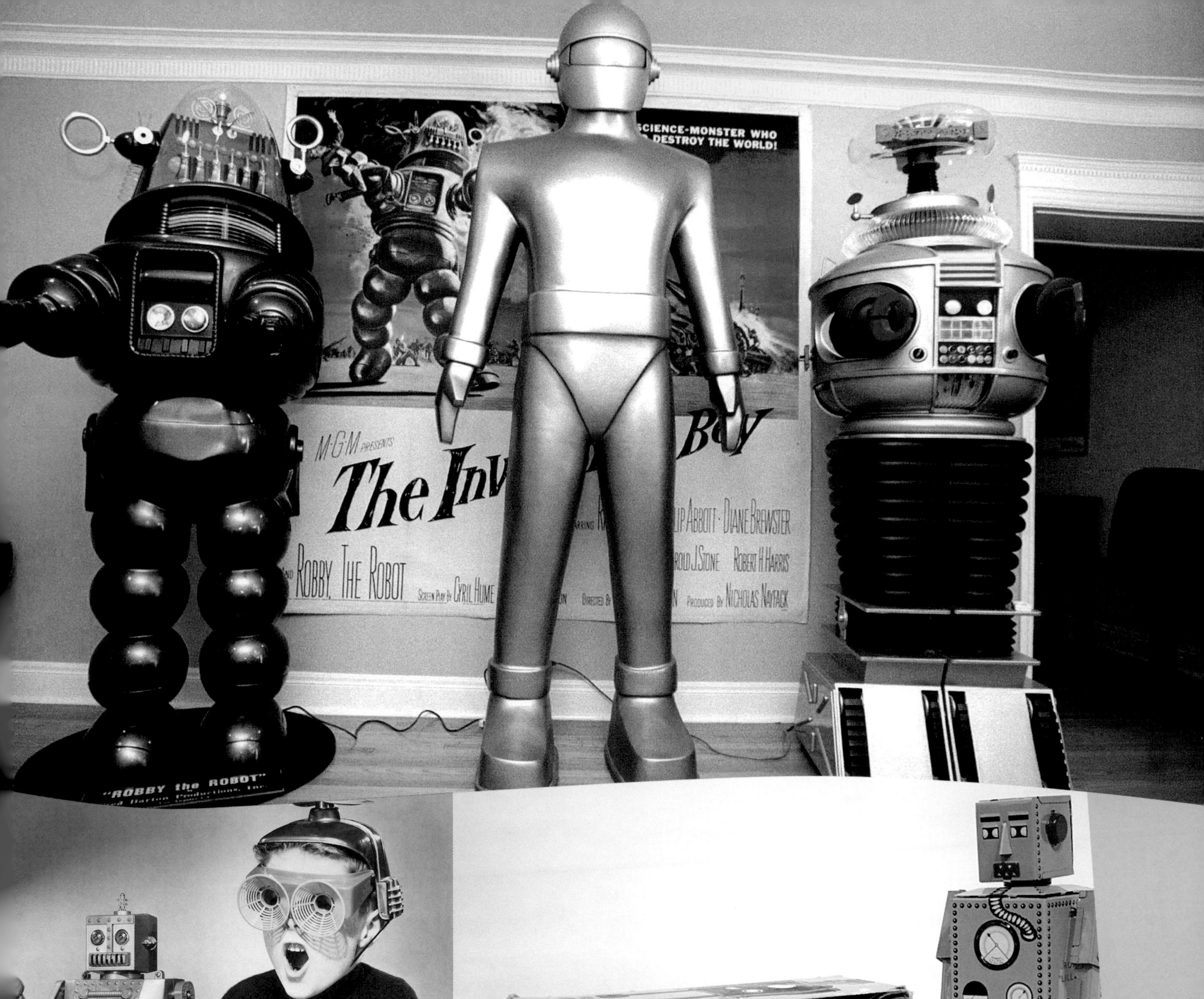
SCIENCE-MONSTER WHO
DESTROY THE WORLD!
M·G·M PRESENTS
The Inv
Boy
LIP ABBOTT · DIANE BREWSTER
ROLD J. STONE
ROBERT H. HARRIS
ROBBY, THE ROBOT
SCREEN PLAY BY CYRIL HUME
PRODUCED BY NICHOLAS NAYFACK
"ROBBY the ROBOT"

Robot Lilliput
ROBOT LILLIPUT

N.P. 5357.

Early toy robots fell into various subcategories. There were Gear or Piston or Sparking robots. Gear robots largely described the robot's basic walking mechanism, while Piston robots sported visible, moving internal mechanisms, and Sparking robots produced some kind of electrical effects, such as flashing eyes. Often, and certainly as the toys became more sophisticated, these loose categories were combined into one. Among the copious toy robots produced in the 50s and early 60s, many—such as Thunder Robot, Robot Sparky, Radical Robot and Winky Robot—are now considered highly collectable and valuable, and are still produced today in detailed and lovingly accurate repro versions. Robot B-9 from the TV series *Lost in Space*, who was himself a re-engineered Robby, enjoyed a Robby-like surge of popularity in the mid-to-late 60s, spawning multiple versions and imitators (such as Rocky).

Most of the early toy robots were sold through U.S. toy manufacturers, their company of origin in Japan either obscured completely or simply hard to trace. Many Japanese robots were built piecemeal, by any number of different companies, often from recycled metal (internal examination might reveal Super Robot or whatever had once been a tin of processed fish). Perhaps the most recognized and since famous of the original Japanese toy companies was Horikawa, but even they were actually just a wholesaler, their product built by the Metal House Company of Tokyo (which still exists today). One of Horikawa's most famous 'classic' robots was the Attacking Martian robot, which had pop-up/light-up chest weapons.

I feel constrained—largely because I had one as a kid and remember it with considerable fondness—to at least mention Magic Robot, which had the distinction of being a game rather than a toy, as such. The robot of Magic Robot stood in the middle of a board littered with questions and answers. When a player stops on a question, he or she has to try and answer it, and having done so (right or wrong) subsequently places the robot, which carried a pointer, in a central magnetized area and he rotated automatically to point out the correct answer. Sure, it was a gimmick, with rather limited play value, but it still had a certain "wow" factor that I remember to this day. The game/quiz, a kind of forerunner of Trivial Pursuit, first appeared in the early 70s. Also fun, and appealing now in a kitsch way, was Rock'em Sock'em Robots,

ABOVE: a radio-controlled version of Robot B-9 from *Lost in Space*, in its original packaging.
OPPOSITE, TOP: box art for Magic Robot, a quiz/board game created and marketed by Merit.
OPPOSITE, BOTTOM: seconds out... the Rock'em Sock'em robots.

a hand-operated boxing game for two players in which you literally tried to knock your opponent's robot's block off. If you connected just right, the robot's head was spring-loaded and would come almost all the way off its shoulders!

By and large, a toy robot was a toy robot. It moved or talked or issued sparks but there was no concept of a backstory or even a brand, something that encouraged/stimulated both the imagination and the acquisitive collector's instinct common in young kids. In Japan, the likes of *Gundam, Macross* and *Robotech*, all of which feature robots to greater or lesser degrees, started as manga (comics) or anime (animation) series and

The fashioning/shaping of Transformers was by no means straightforward

evolved into toylines, but the opposite was unheard of... until *Transformers*. But the fashioning/shaping of *Transformers* was by no means straightforward, involving a strange, circuitous evolution from America to Japan and back again to America.

In the early 1970s, U.S. toy manufacturer Hasbro licensed its immensely popular G.I. Joe action figure line to Japanese company Takara, who took the basic humanoid mould and turned it into a transparent Henshin Cyborg figure, complete with battle armour and accessories. This was subsequently scaled down to become the smaller Microman line, and then (combining two popular concepts, toy cars and toy robots) the revolutionary transforming Diaclone 'Car Robot' series. Finally, in the early 80s, Takara released its New Microman line (with its Microchange sub-series), adding the further refinement of robots that transformed into non-car modes. Then, in 1983, Hasbro re-licensed all the various Takara transforming toy lines and repackaged them as *Transformers* for the U.S. market.

Hasbro took the further step of presenting the largely unnamed cast of characters to Marvel Comics, with the idea of grafting a whole, unifying backstory to the brand. Initially, Marvel's Editor-in-Chief Jim Shooter and then writer/editor Bob Budiansky created a whole world for the toys to inhabit, not simply naming them but giving each character a whole potted profile, detailing rank, role, abilities and so forth. When that was all in place, Hasbro unleashed *Transformers* on an unsuspecting generation of kids, the toy line quickly spinning off into a (Marvel) comic book series and animated TV show. The framing story pitted the heroic Autobots, led by Optimus Prime, against the evil Decepticons, led by Megatron, and told of an ancient civil war on the metal planet of Cybertron and detailed how, more or less by accident, it spreads to present day Earth.

Transformers proved a runaway success for Hasbro, which more or less prompted Takara to apply roughly the same storyline to its subsequent rounds of transforming robots. (The two companies now consult regularly on brand direction and content.) Other companies were quick to see the potential and marketability, as Hasbro had done, of robots that transformed or even those that didn't, as long as they came with attached media, such as comics or animated TV shows. A host of good, bad and indifferent toy robot series followed hotfoot on the success of *Transformers,* and toy stores were inundated by the likes of *GoBots, Starriors, Robo Machines, Voltron* and *Zoids*, to mention but a few. Of these, *GoBots*—which also featured transforming robots—was another Americanized Japanese import (originally called *Machine Robo*) distributed by Tonka. The line was eventually absorbed into Hasbro's *Transformers* brand when they bought out Tonka in 1991.

Zoids, another Japanese/American crossover brand, kept what Transformers and GoBots had done away with—the tiny human or android pilots that controlled the giant robots. Another difference was that the robot toys themselves, which were largely though by no means exclusively modelled on dinosaurs, were often

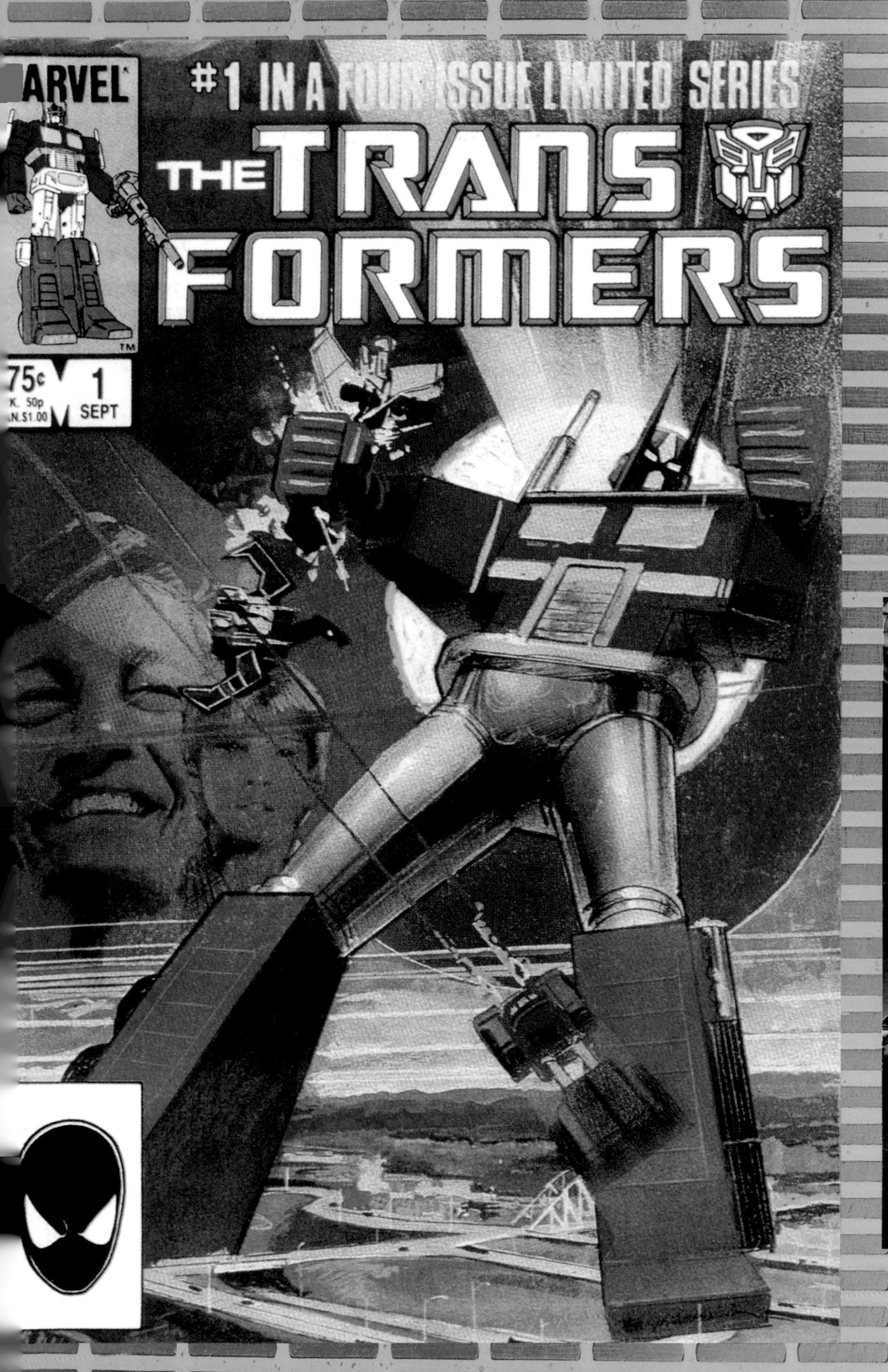

OPPOSITE: a Henshin cyborg, part of an action-figure range developed by Takara. It was the first step on an evolutionary ladder that would lead, ultimately...

LEFT: ... to *Transformers*! Helped by a comic book, animated series...

BELOW: ... and big screen animated movie, *Transformers* would come to dominate the boys' action figure market for the whole of the 80s.

wind-up or battery powered, harking back to the robot toys of yore. A *Zoids* comic was produced in the U.K. but in the U.S. the entire brand was reinvented (again using a Marvel-created backstory) as *Starriors*.

But while transforming and non-transforming robots dominated the toy shelves of the 80s, by the 90s most were gone, replaced by Ninja Turtles and the like. Only *Transformers* kept managing to reinvent itself successfully. What became known latterly as Generation 1 (being the classic 80s Transformers characters and storylines) was succeeded by Generation 2, then *Beast Wars* (in which the cars and jets and tanks were replaced by animal secondary modes) and a number of other variations, right up to the latest: *Transformers Animated*. A live-action movie franchise that kicked off in 2007 continues with *Transformers 2: Revenge of the Fallen* in 2009, both films directed by Michael Bay. Amazingly, *Transformers* is now twenty-five years old and counting...

Over in Japan such longevity for franchises based on robot toys is more commonplace. *Gundam*, which began life in 1979 as *Mobile Suit Gundam*, is still as popular today as it ever was, perhaps more so. Developed primarily by renowned Japanese artist/animator Yoshiyuki Tomino, *Gundam* involves vast, militaristic bi-pedal transforming robots, piloted by humans. Largely, though there are odd exceptions, the machines are non-sentient, but the sheer range and scope of the overtly robotic toys definitely merits a mention here. *Gundam* in Japan, and to an extent worldwide, is a truly massive

phenomenon, with animated movies, TV shows, manga, computer games, toys and more. There's even a massive full-size, 59 feet tall Gundam robot in a special display near Mount Fuji. Likewise, *Macross* (originally Super Dimension Fortress Macross), another tale of intergalactic war waged with giant transforming robots, has nearly 30 years of toy and associated media history in Japan.

Through the 1990s to the present day, robot toys have become increasingly sophisticated, utilizing more in the way of software and machine intelligence programming. Early examples of these toys include Tamagotchi and Furby. A Tamagotchi was essentially a virtual pet, which existed in a hand-held computer module. You could feed or play with your Tamagotchi, but if neglected they could 'die'. Furby, on the other hand, was a cute, owl-like automaton that spoke a unique language of its own, Furbish. The idea was, the more you spoke to them in English the more they 'learned' to speak it back. Furby, created and distributed by American toy manufacturer Tiger Electronics, made its debut in 1998 and quickly became the must-have toy of 1999. Sony's Aibo, a robotic dog initially marketed in 1999 could sense and track a luminescent pink-red ball, allowing it to 'fetch', and could recognize and respond to certain vocal commands. Later models came with an in-built camera and wireless Internet connectivity, allowing them to take and send pictures via email. Sadly, the Aibo line was discontinued in 2006.

OPPOSITE, LEFT: King Gojulas, one of the largest and now rarest of Tomy's *Zoids*.
OPPOSITE, RIGHT: "robots in disguise", a classic selection of Generation 1 Transformers including Optimus Prime, Megatron, Shockwave and Grimlock.
ABOVE: seamless integration of CG robots and live action "Bayhem" in Michael Bay's 2007 summer blockbuster *Transformers*.

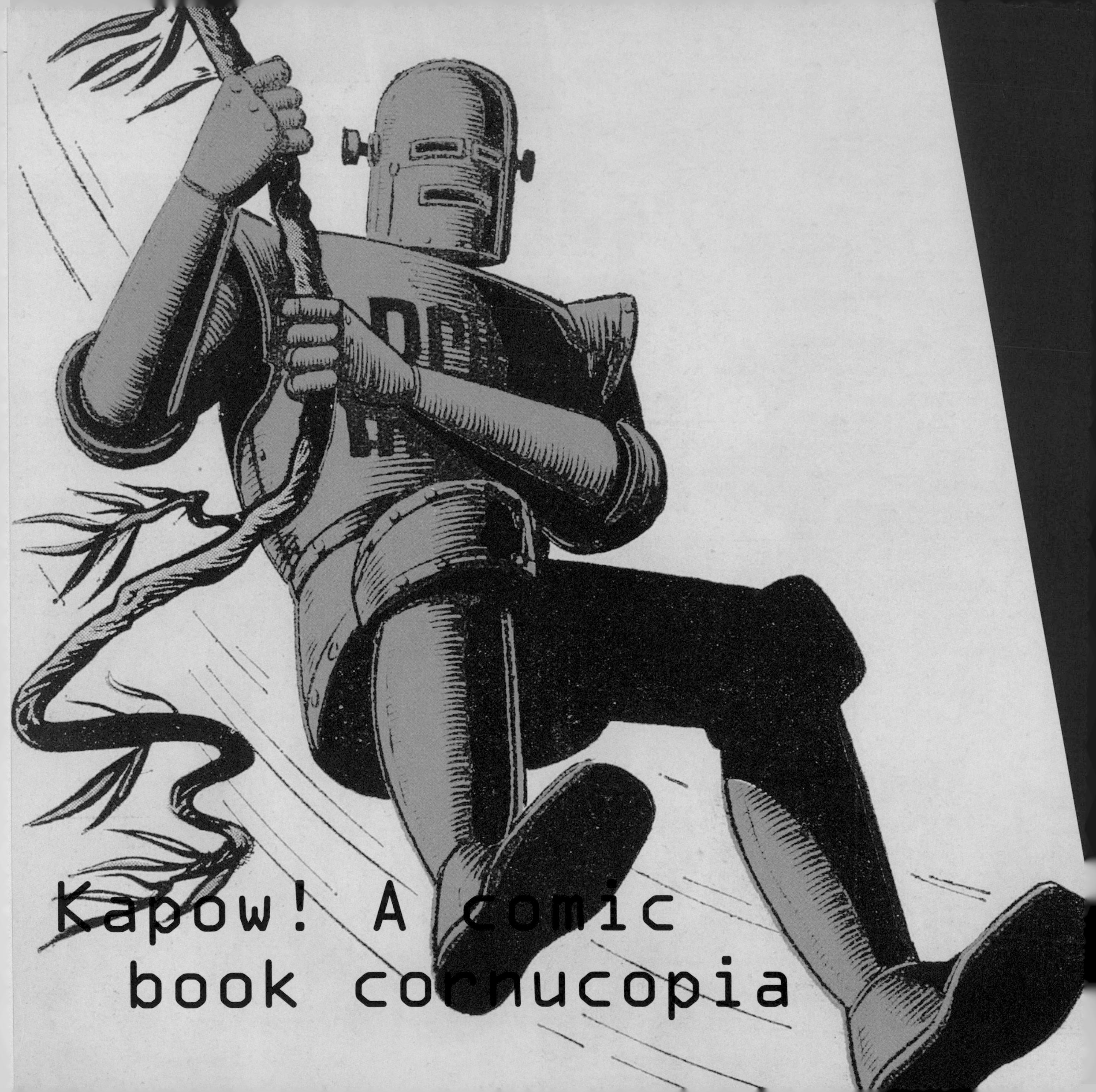
Kapow! A comic
book cornucopia

9

The Mighty World of Robots!

Robots on the whole were interchangeable with 'monster'

Given their shared pulp roots, it was only to be expected that within the often lurid, gaudy covers of kids comic books robots would lurk in numbers.

During the 30s, 40s and 50s, the so-called Golden Age of comics, robots on the whole were interchangeable with 'monster', in that they were generally otherworldly menaces, aberrant creations or mindless tools of arch villains, there to give Superman and his spandex kin something substantial to hit. Though, it should be said that the first ever robot to grace a comic book cover, Bozo the Robot from *Smash Comics* #1 was at least a crime-fighter, albeit one created by an evil genius and subsequently put to better use by fedora wearing hero Hugh Hazard.

The late 50s/early 60s 'monster' books—such as *Tales of Suspense* and *Tales to Astonish*—published by Atlas Comics, the direct forerunner to Marvel Comics, were positively teeming with robots on the rampage. Strangely, though, these early, charmingly naive forays into comics automata often

ABOVE: *Smash Comics* #5, featuring Bozo the Robot. Originally, the strip appeared as 'Hugh Hazard and his Iron Man', later re-titled simply 'Bozo the Robot'.
OPPOSITE: Timely Comics' *Marvel Comics* #1 from 1939, featuring the original (android) Human Torch, as conceived by Carl Burgos.

managed to hit the nail squarely on the head in terms of that pervasive, underlying fear I've alluded to in other chapters. 'Elektro—He Held the World in His Iron Grip!' from *Tales of Suspense* #13 concludes with a sombre sermon on our technological folly: "For to create a machine which can out-think man is to create the instrument of our own destruction." Amen to that. But amidst the alien engines of destruction and Nazi war-machines, to paraphrase many a comic book title or tagline, there came... a hero: the Human Torch.

Not to be confused with the later Silver Age Human Torch, one-fourth of the Fantastic Four, the original Human Torch was an android, the brainchild of inventor Professor Phineas T. Horton. He made his debut in *Marvel Comics* #1, published in 1939 by Timely Comics (the pre-Atlas iteration of Marvel Comics). Able to ignite his entire body, fly and hurl flaming missiles, the Human Torch—occasionally under the alias Jim Hammond—fought alongside fellow Timely heroes Captain America and The Sub-Mariner, often in stories set against a backdrop of World War II. The Human Torch was soon a regular in *Marvel Mystery Comics* (as the title became as of issue #2) and then his own title, *The Human Torch*, until both were cancelled in 1949, as superheroes in general waned in popularity. Latterly, the Human Torch became a core member of the Invaders, a WWII super-

team developed by Roy Thomas in 1969 and he made his first Silver Age appearance in *Fantastic Four Annual* #4, in which he fought the new (human) Human Torch.

The popularity of the original Human Torch seemed to count little in the new Marvel age of comics (which dates roughly from November 1961 when *Fantastic Four* #1 was published), as superheroes made their comeback and robots/androids were relegated once more to the role of general threat (The Living Brain in *Amazing Spider-Man* #9, Dragon Man from *Fantastic Four* #35), disposable henchmen (Doombots, robotic replicas of arch-villain Doctor Doom, from *Fantastic Four* #5 and onwards/upwards), remote-controlled tools (the various Spider-Slayers from *Amazing Spider-Man* #25 et al) and other-worldly menaces (the Kree Sentry from *Fantastic Four* #64, Behemoth from *Tales to Astonish* #79, the Sub-Mariner portion of the shared book). The coldly calculating Mad Thinker merits special recognition here in terms of the sheer volume of wayward artificial creations unleashed on the Marvel Universe, from his Awesome Android in *Fantastic Four* #15 to living computer Quasimodo in *Fantastic Four Annual* #4 and beyond. Only the Recorder, part of a machine race created by the alien Rigellians simply to record unfolding galactic events (first introduced in *Thor* #132) bucked this generally antisocial trend.

Instead it was DC Comics who spearheaded something of a robotic renaissance with *Showcase* #37, which introduced the Metal Men. Created by writer Bob Kanigher and artist Ross Andru, the elemental Metal Men—Gold, Lead, Tin, Mercury and Platinum—were intelligent robots with characteristics,

OPPOSITE: making the transition from Golden Age to Silver Age, the android Human Torch battles his superhero counterpart, Johnny Storm in *Fantastic Four Annual* #4.
LEFT: cover to issue #26 of Gold Key's *Magnus, Robot Fighter*.
BELOW: the Metal Men made their debut in DC's *Showcase* title before moving into their own comic.

both personality and ability-wise, of their namesake metals. Created by Doctor William Magnus, the team fought assorted menaces, often other robot menaces, the Missile Men and the Robot Amazons among them. After a four-issue run in Showcase, the Metal Men moved directly into *Metal Men* #1, headlining a title that ran for fifty-six straight issues. "Doc" Magnus also created the robotic body for the Doom Patrol's Robotman (aka Cliff Steele). But, technically, as his brain is human, transplanted into the body following a high-speed car crash, he doesn't qualify for more than a passing mention.

Robots—while not exactly the headliners—got a more comprehensive comics outing in Gold Key's *Magnus, Robot-Fighter*, the tale

of a human raised by robot 1A in or around the year 4000, who is trained to fight rogue robots. Interestingly, *Magnus, Robot Fighter* uses Asimov's Three Laws of Robotics as its story base, 1A (Magnus' mentor) adhering rigorously to them in his teachings. Artist/writer Russ Manning, who created *Magnus, Robot Fighter*, only actually drew twenty-one (of the forty-six) original issues, the rest being reprints or featuring guest creative teams. The title was resurrected by Valiant (later Acclaim) Comics in 1991 as part of a raft of Gold Key character reinventions.

Meanwhile, in the U.K., comic readers had their own robot heroes, most notably Robot Archie. Archie made his debut in the first issue of boy's adventure comic *Lion* in 1952, in a strip entitled 'The Jungle Robot,' by E. George Cowan and Ted Kearnon, which ran until issue #25, and featured a mouthless, speechless, remote-controlled robot in adventures set largely in the jungles of Africa and South America. He wasn't actually called Archie until 1957, when he returned to *Lion* after an absence of four years in the strip 'Archie the Robot' (later just 'Robot Archie'), now with a mouth and the power of speech. Some years later, Archie became a fully autonomous robot. *Lion* was cancelled in May 1974, but Archie subsequently appeared in Annuals and (in colour) in the short-lived *Vulcan* comic. Versions of Archie would also feature in *Captain Britain* and *Zenith,* until he was finally resurrected as part of Wildstorm's 2005 *Albion* series.

Mytek the Mighty, who made his debut in *Valiant* in 1963, was not simply a robot he was a robot ape, and a giant robot ape to boot! Invented by Professor Boyce, Mytek was created to pacify the warlike African

ABOVE, LEFT: Robot Archie, one of the first and greatest British robot heroes.
ABOVE, RIGHT: robotic ape Mytek the Mighty, from boy's action/adventure comic *Valiant*.
OPPOSITE: *Action Comics* #280, featuring artificial being and walking super-computer Brainiac, one of Superman's most enduring villains.

Askari tribe, who worshipped Mytek as their god. Sadly, this plan went to pot when Boyce's assistant, Gorga, took control of Mytek and then Mytek went and gained his own artificial intelligence and fought Gorga's giant robot and... well, you get the idea. Often in British comics it was as if time had stood still and the strip's depiction of the British in Africa is more akin to 1863 than 1963! Mytek got an update in the 1992 *2000AD Action Special* and also featured briefly in *Albion*. Artist Rich Veitch lampooned the character in his 'Mooretek the Mighty' strip, which featured a giant, hairy Alan Moore (writer of *Watchmen, V For Vendetta* et al).

Other (early) British comics robots included The Steel Commando, a WWII indestructible combat robot who first appeared in *Thunder* #1 (1970), and subsequently in both *Lion* and *Valiant*, Brassneck, a robotic schoolboy with telescoping limbs and neck from *The Dandy* (circa 1964-1968), Klanky, an alien mechanical man from *Sparky* (1966-1974), Smasher, a giant indestructible robot controlled by Doctor Doom (not the Marvel villain) in *Bullet* (circa 1977, later reprinted in *Red Dagger*), Juggernaut, a giant alien robot from 'The Juggernaut from Planet Z' (*Hurricane* #2, 1964 and onwards) and Tin Lizzie (1955) from *The Dandy*, a robot housemaid who fought a long-running battle with robot butler Brassribs.

Evil/villainous robots were, as mentioned previously, plentiful in the worlds of both Marvel and DC Comics during the 60s. But of the mass of largely forgettable mechanoid menaces, a few do stand out and have stood the test of time. Recurring Superman nemesis Brainiac (BRAIN InterActive Construct), who made his debut in *Action Comics* #242, has

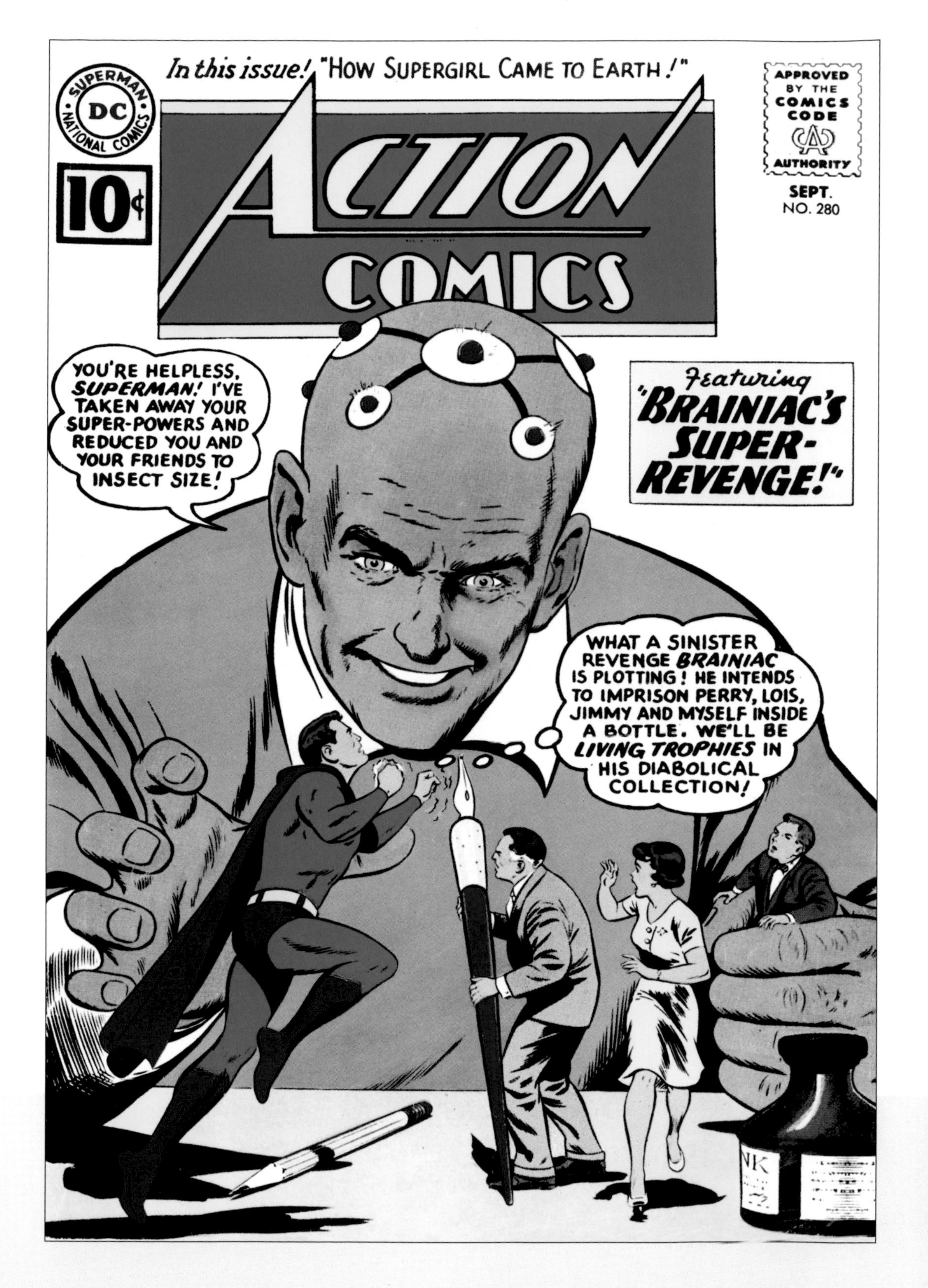

LEFT: the Vision, an artificial being created by mad, bad robot Ultron, made his debut in *Avengers* #57.
OPPOSITE: the mutant-hunting robotic Sentinels have plagued the X-Men, on and off, for over 40 years.

been through many versions and revisions as DC tweaked at its own often-convoluted continuity, but originally he was a machine being created by the Computer Tyrants of Colu. Brainiac's appearance may have changed drastically over the years, from green-skinned with red electric terminals across his forehead to all-metal skeleton with visible brain (inside a honeycomb-style brain case) to almost totally robotic (as Brainiac-13), but he remains one of Superman's top three villains. Brainiac has appeared in almost every animated Superman series as well as the live-action *Smallville* (played by James Marsters).

And where DC had Brainiac, Marvel had Ultron, a self-upgrading, evolving machine entity of its own. Before we even meet Ultron (as the Crimson Cowl in *Avengers* #54), he's already upgraded to Ultron-5 and brainwashed his inventor, scientist Henry 'Hank' Pym, to make him forget he ever created him. At the last count, the character had reached Ultron-19 in terms of his many and varied upgrades and incarnations. But while Ultron, with his almost indestructible adamantium armour and pathological, Oedipus-style hatred of his 'father', Hank Pym, was a fine and enduring presence in his own right, he is also the roundabout progenitor of two more of Marvel's robotic cast of characters. Ultron's second assault on the Avengers involved the creation of the artificial being (synthezoid) known as The

Vision, who would go on to rebel against Ultron and become a fully fledged member of the Avengers. Ultron also coerces Hank Pym into building him a robotic bride. In time, this latter creation, dubbed Jocasta, also sides with the Marvel heroes.

The mutant-hunting Sentinels and their supreme intelligence, the Master Mold, carved out a substantial bad robot niche of their own. Introduced in *X-Men* #14–16, the Sentinels were the brainchild of Dr. Bolivar Trask, who programmed them to hunt, capture and even kill mutants. However, the Sentinels soon concluded they knew best and set out to take over the entire world. Since then, the Sentinels have returned time and time again to plague the X-Men and the world in general, and in a possible future revealed in the story 'Days of Future Past', in *Uncanny X-Men* #141–142, they control much of the planet, having extended their mutant-hunting remit to include any super-powered individual. While huge in stature and utterly unswerving in their task, only a few Sentinels have actually possessed real self-awareness.

The Vision aside, robotic or android heroes were a rare beast in the Marvel universe. DC, in addition to the Metal Men, had their own Vision-equivalent hero in the shape of the Red Tornado, an artificial being created by the villain T. O. Morrow in order to infiltrate and destroy the J.S.A. (Justice Society of America). The Red Tornado first appeared in *Justice League of America* #64, though the elemental being had previously appeared in *Mystery in Space* #61. Finally, in 1977, Marvel

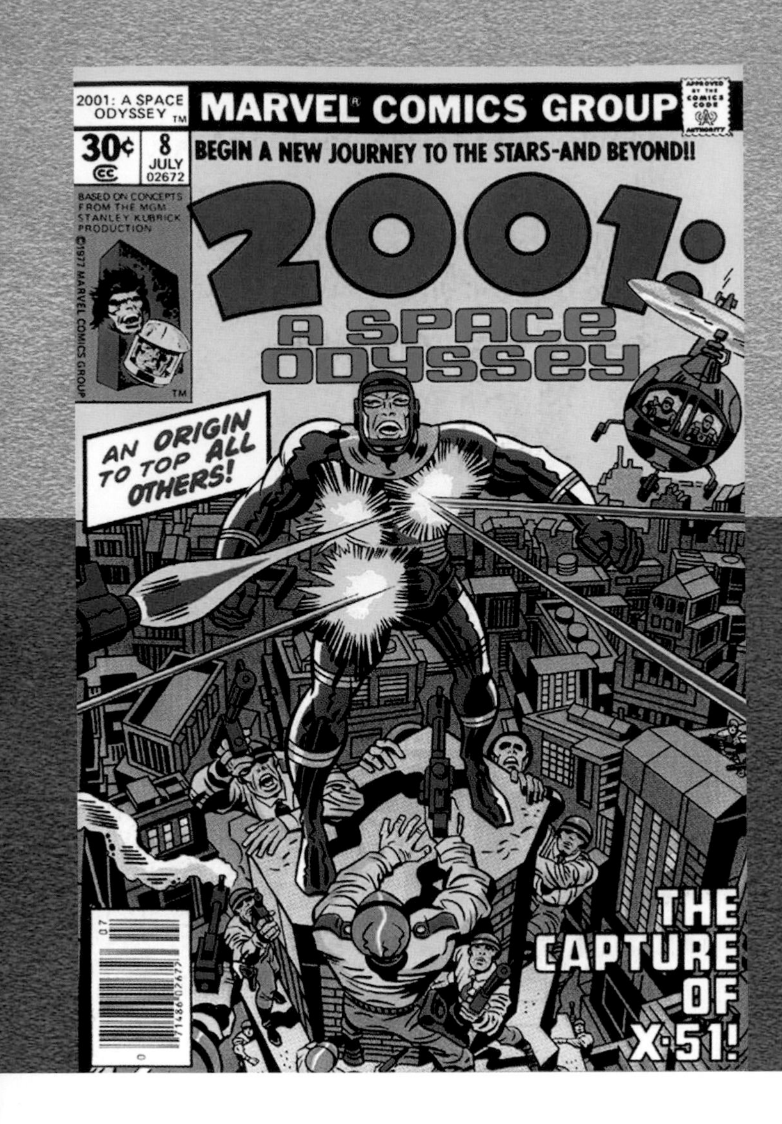

unveiled its first robotic title character—Machine Man.

X-51 or Mister Machine, as he was originally called, first appeared in Marvel's *2001: A Space Odyssey* spin-off comic series, and his origin involved the artificial being coming into contact with one of the monoliths from the book/film of the same name. However, by the time *Machine Man* #1 hit in 1978, the monolith element had been dropped from his rehashed origin story, presumably as the rights to those extended only as far as the 10-issue *2001: A Space Odyssey series*.

Created by the legendary Jack Kirby, X-51 is a machine lifeform developed by Dr. Abel Stack for the U.S. army that eventually assumes the human name Aaron Stack as it/he attempts to fit in with humankind. Machine Man's solo run lasted just nine issues, but after an appearance in *The Incredible Hulk* the series was extended to nineteen. A later mini-series, written by Tom DeFalco and drawn by Barry Windsor-Smith moved the action to the year 2020, giving Machine Man a cyberpunk twist.

Back in the U.K. comic book robots received a radical makeover with the advent of counter-

ABOVE, LEFT: Machine Man's first appearance (as X-51) came in issue #8 of Marvel's short-lived *2001: A Space Odyssey* spin-off comic.
ABOVE, RIGHT: a few evolutions onwards, Machine Man made a real impact thanks to Tom DeFalco and Barry Windsor-Smith's reinvention.
OPPOSITE: cover to *2000AD* Prog #125, featuring the A.B.C. Warriors. Art by Mike McMahon.

culture sci-fi comic *2000AD.* Outdated British comic traditions were gleefully trampled underfoot by a range of psychotic, punk-era, in-your-face mechs with a surfeit of attitude. These anarchic leanings were personified by the A.B.C. Warriors (Atomic Bacterial Chemical), a robot combat unit populated by the unsavoury likes of Hammerstein, Mek-Quake, Ro-Jaws, Joe Pineapple and others, all borderline sociopaths with big guns and vast capacity for bloody carnage. The characters Ro-Jaws and Hammerstein first featured in the 'Ro-Busters' strip, which came to *2000AD* in Prog #86 via *Starlord* comic. 'A.B.C. Warriors' itself kicked off in Prog #119, following the war-droids through the final stages of the epic Volgan War, also previously established in 'Ro-Busters'.

Other notable *2000AD* robots include the entire supporting cast of 'Sam Slade, Robo-Hunter,' in which the eponymous Sam, a Bogart modelled private eye pastiche who hunts/terminates rogue robots, is joined by the likes of Cutie, inept assistant Hoagy and—due to an embargo on characters smoking in the comic—Stogie, a robot cigar. Also, Armoured Gideon from 'Armoured Gideon', a demon-hunting robot created by a race known as the Silent Ones, and Walter the Robot (or "Wobot", due to the character's lisp), put upon, adoring, neglected domestic servant of Judge Dredd, in many episodes of 'Judge Dredd' itself and his own one-page 'Walter the Robot' strip.

In fact, robots are omnipresent in the future world of Judge Dredd and there have been at least two all-out robot uprisings in the strip's history. And even outside of these core strips, in the fictional editorial world of *2000AD* all the writers, artists, letterers and colourists are droids (art-droid, script-droid, etc), many of whom have been visualized in the comic itself.

But not all 'bots with attitude came from the pages of *2000AD*. Death's Head, a robotic freelance peacekeeping agent (don't call him bounty-hunter!) with a decidedly pragmatic approach to his work emerged from Marvel UK's brief foray into American format comic books in the mid-80s. Originally presented in a one-page strip, Death's Head moved via *Transformers* (U.K.), *Doctor Who Monthly* and *Dragon's Claws* (another U.S. format Marvel U.K. title) into *Death's Head* #1. A further nine issues followed, plus a graphic novel 'The Body in Question', and various guest-shots in Marvel U.S. titles. Death's Head was re-imagined as Death's Head II in the 90s and, most recently, as Death's Head 3.0 (in the pages of the 2005 version of *Amazing Fantasy*).

Naturally, Japan has its own comic book (manga) robots, including Osamu Tezuka's *Astro Boy* and Masamune Shirow's *Ghost in the Shell*. Astro Boy was the creation of robotics engineer Dr. Boynton. Having lost his son in a tragic car accident, Boynton attempts to recreate him in robot form but ultimately abandons his creation, who is then taken in by kindly Android's Rights activist Dr. Elefun, who trains Astro Boy and guides him in the use of his special powers. Astro Boy made his manga debut in 1952, and has since become one of the most instantly recognizable Japanese manga/anime characters in the west. In *Ghost in the Shell* the line between organic and technological is blurred to the point where terms such as human, cyborg and robot become almost irrelevant. Special agent Motoko Kusanagi, herself a largely cybernetic being, investigates a new A.I. known as 'the puppeteer' that can possess other A.I.s. *Ghost in the Shell* has spawned two manga sequels, three animated movies and a TV series.

There are, quite simply, far too many robot characters in comics to provide a definitive list here, but others I feel deserve at least a mention include Spartan, an artificial being from the WildC.A.T.s super-team created by Jim Lee, Amazo, a villainous artificial being programmed with the various abilities of the Justice League of America members, H.E.R.B.I.E (Humanoid Experimental Robot, B-type, Integrated Electronics), the Fantastic Four's 'cute' robotic assistant from both cartoon series and comic, Ghost Rider 2099, the technological descendent of the Marvel supernatural anti-hero, Skeets, the companion of Booster Gold, Torgo, the gladiatorial alien from the classic, in my opinion, *Fantastic Four* #91–93 (though it's never overtly stated he is purely a robot), Rusty the Boy Robot from Frank Miller and Geof Darrow's *The Big Guy and Rusty the Boy Robot* and *Arsenal*, a robot bruiser from *Iron Man* and *The Avengers*.

ABOVE: whatever you do, don't call him bounty hunter! Marvel UK's, er, freelance peacekeeping agent, Death's Head, from *Death's Head* #1.
OPPOSITE: dramatic promotional art for the 1995 animated feature film based on Masamune Shiro's manga *Ghost in the Shell*.

10

The Shape of Robots to Come

Science fiction
meets science fact

What next? I guess that's the big question. Where does the science of (real) robots go from here? How far are we from the fictionalized ideal (or its negative flipside) of reactive, learning, fully functional robots in everyday life?

Judging by where we are now, you'd have to say it's all a long way off, and the mass produced servile robot will remain the domain of science fiction/fantasy for some good while yet. But you can never tell.

A simpler though no less intriguing question to ask and maybe answer is: when does a robot stop being a toy and become a piece of real technology? In *Robot Wars,* a TV show developed in America by Marc Thorpe and licensed to the U.K., remote-controlled robots with names such as Sgt. Bash and Sir Killalot are pitted against one another in vaguely gladiatorial duels or other trials that test endurance or versatility. One might argue that these robots are a stage up from toys, but maybe not quite cutting edge technology either. So where does the one become the other?

Robosapien, by its very marketing, falls into the toy category, but what about, say, Asimo – toy or technology? For the longest time, the problem robotics engineers wrestled with was how to get a robot to walk like us. Sounds like it should be straightforward, but it's not. We walk off balance, our centre of gravity slightly off the perpendicular, providing us with extra momentum from step to step. We don't even think consciously about it, but every tiny thing a robot does has to be painstakingly taught. Teaching, or rather programming a robot to walk in a smooth, continuous gait without falling over has proved difficult in the extreme. Many budding roboticists failed miserably, at least until Asimo came along. Asimo, though there many variations and different sophistications of Honda's pint-sized humanoid robot, analyses–at incredible speed–each subsequent small motion before it makes it, allowing it to replicate a steady, human-ish walk. It can even run and climb stairs.

Asimo (Advanced Step in Innovative MObility, not–its makers stress–a slight contraction of Asimov) was first unveiled in 1986, in prototype form, and then in 1993 in humanoid form. Its more current and familiar incarnation, a sort of Munchkin astronaut, possesses other features too. It can recognize and identify objects, gestures and faces, it can map and navigate its immediate environment and even distinguish sounds. But, ultimately, can it think? Can it reason? Scientists in Germany believe so, having demonstrated that their Asimo can weigh up two different versions of the same household item, such as chair, and make a judgement call as to whether which, if either, falls into its broad definition of what makes a chair.

BELOW: the aftermath of *Robot Wars*, a U.S. TV series (a version of which appeared in the U.K.) in which remote-controlled robots are pitted against one another in the arena.

LEFT: credit where credit's due. Honda's Asimo pays tribute to Karel Capek, author of *R.U.R.* **OPPOSITE:** light me up! Elektro (built by Westinghouse), star of the 1939 World's Fair, gives his opinion on the smoking ban. TV singer Jeanne Dowling obliges.

But in the end, as sophisticated as it may be, Asimo is still just a series of inputted programs, and cannot react to anything not already experienced without a new set of programs and prompts. It's telling that Asimo is a featured attraction at Disneyland, in its 'Tomorrowland' exhibit. In the final analysis, Asimo may be little more than a sophisticated sideshow attraction, especially as viewed from our own Tomorrowland. Are we really any much further forward than Elektro, unveiled at the 1939 New York World's Fair, with its 78-rpm disk of pre-recorded vocal one-liners and cigarette smoking gimmicks?

Undaunted by any such perceived lack of real progress scientists continue to churn out variations on an Asimo theme. There's CBi (Computational Brain-interface), which—according to the press release from the Japan Science and Technology Agency is the "world's first bipedal locomotion with a humanoid robot controlled by cortical ensemble activity with a real-time network brain interface." Phew. Which boils down to a robot that can actively mimic human behaviour. CBi's creators claim to be able to input programming data (via the Internet) based on real-time human or animal brainwave patterns recorded during a given action or set of actions. This seems, and perhaps is, groundbreaking, but CBi is still effectively a very clever mimic. On a more straightforward level, CBi can also follow

Are we
really
any much
further
forward
than
Elektro?

RIGHT: teaser poster for 2009's blockbuster sequel, *Transformers: Revenge of the Fallen.* The first film took over $700 million worldwide. **OPPOSITE:** Keanu Reeves steps into Michael Rennie's shoes (and has trouble filling them) in the recent (redundant) remake reimagination of *The Day the Earth Stood Still.*

and need? Sadly, the answer is no. Or, at least, not much! But while real world robots may still be in their creative infancy, the imaginations of writers and filmmakers show no signs of letting the evolutionary pace of robots slow. *The Terminator* franchise has embarked on a new trilogy of movies, with Christian Bale as John Connor and—as the first film is set largely at the height of the war between man and machine—no doubt a whole slew of new Skynet Terminator robots and hunter-killers. *Transformers* rolls on also, with *Transformers: Revenge of the Fallen*, with its ever-expanding cast of giant transforming robots. And as recent movies such as *Eagle Eye* and the remake of *The Day the Earth Stood Still* show, robots and A.I.s are still a big box office draw.

One thing's for sure, we don't have to start worrying about robot uprisings or wayward A.I.s sparking nuclear apocalypse. At least... not yet! But always keep the fear. Did I mention? Fear is good.

KEANU REEVES

JENNIFER CONNELLY

THE DAY THE EARTH STOOD STILL

IN

IMAX

Acknowledgements

Rad Robots Picture Credits

AA: Advertising Archives
AKG: AKG-images
MSC: Movie Store Collection
MEPL: Mary Evans Picture Library
KC: Kobal Collection
GI: Getty Images
RF: Rex Features
VMA: VinMag Archives

t = top
b = bottom
m = middle
l = left
r = right

Introduction

Page 5 Jake Tilson; 6 (and 120) AA; 9 Marvel Comics; 10 MGM/MSC; 11 C-2 Pictures/WB TV/KC/Desmond, Michael; 13 20th Century Fox/KC/Digital Domain.

Chapter One: Rise of the Robots

Page 14 (and (b) 22) UFA/KC; 16 (tl and tr) Ben Sweeney/www.leonardoshands.com (b) from *Leonardo's Lost Robots* by Mark Elling Rosheim; 17 Flemish School/GI; 19 Columbia/KC; 20 MEPL/Alamy; 21 (t) MEPL/Alamy (b) Dragon News/RF; 22 AA; 23 Universal/KC (b) MEPL/Alamy; 24 John Mustain/Stanford Special Collections and University Archives; 25 Lebrecht Music and Arts Photo Library/Alamy; 26 MEPL/Alamy; 27 BBC/Corbis.

Chapter 2: Robots go Galactic!

Page 29 (and 37) United Artists/KC; 31 Boris Bilinsky/AA; 32 U.F.A/Album/AKG; 33 (t) UFA/KC (b) UFA/BFI Stills; 34 Pictorial Press Ltd/Alamy; 35 20th Century Fox/KC; 36 (l) 3-Dimensional Pictures/KC (r) Republic/KC; 38 MGM/Aquarius; 39 MGM/KC; 40 MGM/KC; 41 Paramount TV/KC; 42 Paramount/KC; 43 Pictorial Press Ltd/Alamy.

Chapter 3: Asimov's Robots

Page 45 (and 55) Ladd Company/BFI Stills; 46 VMA; 47 (tl) SNAP/RF (bl) MEPL/Alamy (r) MEPL/Alamy; 48 AA; 50 20th Century Fox/KC/Digital Domain; 52 (t) Columbia/Aquarius (b) BBC; 53 Genie Productions/KC; 54 (t) Ladd Company/BFI Stills (b) Panther Books Ltd; 56 BBC/Nigel Robertson; 57 Warner Brothers/KC.

Chapter 4: When Robots go Bad

Page 59 (and (t) 72) Everett Collection/RF; 61 Aardman Animations Ltd/Wallace & Gromit 1995; 62 MGM/KC; 63 MGM/BFI Stills; 64 (tl) AIP/Aquarius (br) Hemdale Film Corporation/BFI Stills; 65 Orion/KC; 66 Carolco/KC; 67 (t) IMF 3/ILM/Album/AKG (b) Walt Disney/KC; 68 ITV/RF; 69 Orion/KC/Newcomb, Deana; 70 Tri-Star/KC; 71 Concorde Pictures/Trinity Pictures/KC; 72 (b) VMA; 73 BBC/William Baker.

Chapter 5: Robots with Dangerous Curves

Page 75 (and 84) Amblin/Dreamworks/WB/KC/James, David; 77 (t) New Line/KC (b) Hulton Archive/GI; 78 Columbia/KC; 79 Ladd Company/Warner Bros/Album/AKG; 80 Universal/Everett/RF; 81 (l) VMA (r) Photos 12/Alamy; 82 (t and b) VMA; 85 C-2 Pictures/MSC; 86 (t) Tribune/Everett/RF (b) Universal/Everett/RF; 87 Sci-Fi Channel/KC; 88 (t) NBCUPHOTOBANK/RF (b) Everett Collection/RF; 89 Sorayama/Artspace/Uptight, 2009 www.sorayama.net

Chapter 6: Spaced out Robots!

Page 90 (and 103) MGM/UA/KC; 93 Copyright Control/Aquarius; 94 Granada Publishing Ltd; 95 VMA; 96 United Artists/KC; 97 (l) Universal/KC (r) MGM/KC; 98 IMF 3/Zuckerman, Robert/Album/AKG; 99 (t) VMA (b) MGM/BFI Stills; 100 Jack H Harris Enterprises/KC; 101 Disney/BFI Stills; 102 VMA; 105 Warner Bros/BFI Stills.

Chapter 7: Robots with a Twist

Page 107 (and (b) 119) VMA; 109 (t) Lucasfilm/KC/Hamshere, Keith (b) Lucasfilm/20th Century Fox/ALB/AKG; 110 VMA; 111 Tri-Star/KC; 112 Walt Disney/KC; 113 Universal/Everett/RF; 114 BuenaVista/Everett/RF; 115 MGM/KC; 116 Universal TV/KC; 117 (t) Universal TV/KC (b) BBC/Aquarius; 118 20th Century Fox/Everett/RF; 119 (t) Southern Television/BFI Stills; 120 AA; 121 VMA.

Chapter 8: Toy Robots

Page 123 (and (bl) 125) Bettmann/Corbis; 124 Robert Evans/Alamy; 125 (t) GI (br) www.rogerarts.com; 126 Ron Browne/www.theoldrobots.com; 127 (t) Trevor Smith/Alamy (b) Daniel Hurst/Editorial/Alamy; 128 www.fanmode.net; 129 (l) Marvel Comics (r) Deg/Hasbro/Marvel/KC; 130 (tl) Alex Bickmore/www.toyarchive.com (tr, m, b) Paul Cannon; 131 Dreamworks/KC; 132 (t) Roslan Rahman/GI (b) Chris Wilson/Alamy; 133 RF.

Chapter 9: The Mighty World of Robots!

Page 134 (and (l) 140) VMA; 136 Golden Age Comics; 137 Marvel Comics; 138 Marvel Comics; 139 (l) Western Publishing (r) DC Comics; 140 VMA; 141 DC Comics; 142 Marvel Comics; 143 Marvel Comics; 144 (l and r) Marvel Comics; 145 Rebellion A/S www.2000adonline.com; 146 Marvel Comics; 147 Manga Entertainment/Album/AKG.

Chapter 10: The Shape of Robots to Come

Page 149 (and (r) 155) Barcroft Media; 151 Spike TV/Everett/RF; 152 AFP/GI; 153 Bettmann/Corbis; 154 (t) Paolo Patrizi/Alamy (b) Frances M. Roberts/Alamy; 155 (l) Scott Nelson/Stringer/GI; 156 Paramount Pictures/Album/AKG; 157 Twentieth Century-Fox Film Corporation/KC.

The author and publishers would like to thank the following for kindly supplying images to use in this book:

Polly Armstrong and John Mustain at Stanford University Libraries
Alex Bickmore
Ron Browne
Paul Cannon
Rufus Dayglo
Gobi at fanmode.net
Roger at rogerarts.com
Mark Rosheim
Ben Sweeney
Miharu Yamomoto

Every effort has been made to contact t owners of photographs used in this book but if any credit has been inadvertently omitted, please contact the publishers fc inclusion in future editions.

Typography

Typeset in Denedo by Carlos Fabián Camargo Guerrero, 2001-2007. Interstat designed by Tobias Frere-Jones, 1994. C Nine designed by Astrid Scheuerhorst, 2000. OCR-A designed in 1968 for American Type Founders – OCR-A is a fo to be read by machines.